Creative Thinking Through Mathematics

This book engages students in interactive explorations that encourage creative thinking in their approaches to mathematics. Organized around five themes, the fifteen quests invite students to notice, wonder, and draw upon their mathematical reasoning as they connect mathematics to other disciplines and everyday situations. The quests foster students' creativity by posing open-ended questions, encouraging reflection, inviting playfulness and curiosity, and connecting to research-based traits of creativity. The authors broaden the notion of what it means to be a mathematical thinker, provide access to mathematics through creative thinking, and celebrate the idea that everyone has the capacity to explore mathematics. Each chapter concludes with a project idea that encourages students to personalize their learning around the chapter's theme. Accompanying worksheets and other resources are also available online. This book is written for undergraduate students, including preservice teachers, and is suitable as a textbook for general education mathematics courses. It can also be used within teacher professional learning communities and professional development programs as a supplement for more specialized mathematics education courses, or anywhere there is interest in approaching mathematics creatively.

Mika Munakata is Professor of Mathematics Education at Montclair State University, New Jersey. She teaches mathematics, research, and pedagogy courses to preservice and in-service mathematics teachers as well as mathematics courses for non-STEM majors.

Eliza Leszczyński is Assistant Teaching Professor at Montclair State University, New Jersey. She teaches mathematics courses for STEM majors, general education mathematics courses, and mathematics education courses for preservice teachers.

Creative Thinking Through Mathematics

A Student's Guide to Mathematical Explorations

Mika Munakata and Eliza Leszczyński

Routledge
Taylor & Francis Group
NEW YORK AND LONDON

Designed cover image: © Getty Images

First published 2026
by Routledge
605 Third Avenue, New York, NY 10158

and by Routledge
4 Park Square, Milton Park, Abingdon, Oxon, OX14 4RN

Routledge is an imprint of the Taylor & Francis Group, an informa business

ISBN: 978-1-032-89080-7 (hbk)
ISBN: 978-1-032-89079-1 (pbk)
ISBN: 978-1-003-54109-7 (ebk)

DOI: 10.4324/9781003541097

Typeset in Optima
by Apex CoVantage, LLC

Access the Support Material: www.routledge.com/9781032890791

Contents

About the authors

Mika Munakata is a professor of mathematics education at Montclair State University, New Jersey, where she teaches mathematics, research, and pedagogy courses to preservice and in-service mathematics teachers as well as mathematics courses for non-STEM majors. Her research areas are creativity in STEM, teacher leadership, and undergraduate education. She is involved in professional development projects in mathematics and science in partnership with local districts.

Eliza Leszczyński is an assistant teaching professor at Montclair State University, New Jersey. She teaches a wide range of mathematics courses, including courses designed for STEM majors, general education mathematics courses, and mathematics education courses for preservice teachers. She is committed to helping every student feel like they truly belong in mathematics and in the wider university community.

Preface: A note to instructors

Background and motivation

The idea for this book originated with our project, "Engaged Learning Through Creativity in Mathematics and Science", funded by the United States National Science Foundation (NSF Award #1611876; Montclair State University, NJ, USA; 2016–2021). As part of the project, we developed instructional modules for an undergraduate general education mathematics course that guide students on explorations of mathematics through the lens of creativity. The contents of this book are based on many of those modules: All of the activities and projects we present have been implemented in our undergraduate mathematics courses in one form or another.

Students in mathematics classes often question the utility of mathematics by asking "When am I ever going to use this?" While it is important to make mathematics relevant to students' lives and interests, this book aims to expand the ways in which we consider mathematics useful. We encourage explorations of mathematics for mathematics' sake, with the hope that students will explore mathematical ideas in creative ways so that their new *ways of thinking* become relevant to their lives. This is not to say that the mathematics content we explore in this book is not 'useful'. In fact, we present ideas that are foundational to topology, modeling, and data science, just to name a few contemporary areas of mathematics. Rather, we recognize that we must also give at least equal attention and value to the process involved in one's approaches to mathematics. We believe that creativity should be central to that approach.

This book is intended primarily for undergraduate mathematics courses for students not pursuing degrees in science, technology, engineering, or mathematics (STEM). The book is aligned with current considerations for the general education curriculum proposed by the American Association of Colleges and Universities (Hanstedt, 2024) and includes recommendations for undergraduate mathematics courses endorsed by the Mathematical Association of America (Saxe & Braddy, 2015) and the American Mathematical Association of Two-Year Colleges (2018).

General education mathematics courses are often designed to expose students to the beauty of mathematics and to help them develop an appreciation for the mathematics around them. Students in these courses come from a variety of backgrounds and have an equally diverse set of educational and career goals. We would be delighted if this book encouraged students to pursue more mathematics courses. But for those that continue on to non-STEM majors, the general education mathematics course may be the last or only college mathematics course that they take. Therefore, it is important that it makes a lasting impact – by presenting mathematics as a way of thinking that is accessible to all students.

By introducing mathematics topics that are likely unfamiliar to these students and encouraging them to explore those topics in new ways, we attempt to level the playing field. Every student interacting with this book has the opportunity to start fresh and is able to access mathematics on their own terms. We provide various access points to mathematics to help students explore their own potential for creative thinking within the context of mathematics. Our intention is to enable students to activate their prior knowledge without over-directing or over-shadowing their explorations and discoveries. Many explorations in the book foster creativity through connections of mathematics to other areas of inquiry – including writing, art, science, photography, and geography – to promote engagement with mathematics in a variety of contexts using a diverse set of tools. We hope that students of various interests, goals, and backgrounds will use this book to engage with mathematics in their own personal way. Finally, because so many of our explorations are, in fact, based on problems that are unsolved or up for debate (what is the best map?) or have multiple possible solutions (what is a good estimate?), we are hoping to instill in students an appreciation for creative and reflective explorations that may or may not lead to satisfying solutions.

Our approach to mathematics through the lens of creativity is perhaps especially timely given the advances in technology we are now experiencing. Technology enhances how we learn and communicate – from performing calculations quickly to producing elaborate graphics. We rely on technology to compute, research, and create models. Even with the affordances that technology provides, however, it is good to be mindful of the benefits of 'hands-on' learning. Google searches or artificial intelligence cannot help us experience making links out of strips of paper to support geometric reasoning or tossing a beachball to investigate probabilistic reasoning. By making our explorations open to students' multiple approaches and perspectives, we are hoping that students are able to personalize their learning of mathematics and to experience making sense of the content in their own ways. Technology can support this kind of learning, but it cannot replace it: Students will need to do the explorations on their own or in collaboration with classmates.

We present this book as an alternative to standard introductory college textbooks for non-STEM majors. Books written for this audience often *re*-present familiar topics from K12 mathematics. In this book, we present mathematics that we suspect is unfamiliar to most college students and encourage explorations through new approaches. As we mentioned, although some students might decide to study mathematics further, the general education mathematics course may be the last formal mathematics class students take; therefore, we consider this course to be our last chance to expose students to the beauty of mathematics and to challenge them to see the many ways in which creativity and mathematics are connected. This textbook was written to support this endeavor. It is our hope that this appreciation for the role of creative thinking in mathematics enhances the way students come to view their world.

Creativity and mathematics

The need to foster creativity in the undergraduate curriculum has been identified and advocated for in the literature on undergraduate mathematics teaching and learning (e.g., Regier & Savić, 2020; Savić et al., 2022). In particular, studies focusing on *teaching* for creativity have emphasized the importance of implementing teaching practices and tasks to elicit students' mathematical creativity (e.g., El Turkey et al., 2020; Munakata et al., 2021; Sternberg, 2019). Research on *learning* through creativity has elucidated students' perspectives on the interplay between creativity and mathematics (Cilli-Turner et al., 2023), further informing our approaches to teaching mathematics. Our book challenges both students and instructors to embrace creativity in mathematics by noticing, wondering, conjecturing, and investigating mathematical contexts and ideas.

Creativity is domain-specific (Runco, 2004), meaning that, for example, creativity in the arts can entail different processes than creativity in mathematics. Creativity in mathematics involves engaging in non-algorithmic decision-making (Ervynck, 1991), generating novel solutions to problems (Chamberlin & Moon, 2005), being flexible (Lithner, 2008), and moving beyond typical recall questions (Sternberg, 2010, 2015). By exposing students to questions and ponderings that are intended to stretch their imagination and expand their notions about mathematics, we aim to encourage approaches and decisions that feel original to them. Our hope is that students can then begin to see mathematics in a new light and come to see themselves as being creative as they explore mathematics.

Two general criteria have been proposed to assess whether or not something is creative: novelty and usefulness (Runco & Jaeger, 2012). With regard to the first criterion, we expect that students' ideas will often be novel to them. To this end, we ask students to embrace different approaches (including unsuccessful ones) and reflect upon their learning, especially as it relates to their new approaches and new views of mathematics. As for the latter requirement, although usefulness could refer to usefulness in 'real' life, we apply it here more specifically to mathematics, where the solution or process must make sense within the context of the problem at hand. In other words, it's great to be different, but the solution and the justification for it also must make sense mathematically. Because the creative thinking we are encouraging is grounded in the context of specific mathematics content, students are asked to justify their thinking and make connections using mathematical language. Students are tasked with making the case that their ideas and approaches are mathematically meaningful and appropriate.

There are also, of course, different orders of creativity. In fact, researchers have identified the four C's of creativity, ranging from 'mini-c' through 'little-c' to 'Pro-c' and 'Big-C' (Beghetto & Kaufman, 2007; Kaufman & Beghetto, 2009; Sternberg, 2019). Creativity with a 'Big-C' is associated with major breakthroughs: One develops a novel gene-editing technique or introduces a new genre of music, for example. These acts often have enduring, far-reaching impact. 'Pro-c' refers to creativity that relies on expertise and has influence primarily within one's professional circle, such as the development of an effective marketing campaign. When one produces a work that is significant to oneself but has a smaller reach, such as a poem they write for their child or a sweater they knit for their dog, they are engaging in 'little-c'. Finally, creativity that transforms one's own perspective or learning is referred to as 'mini-c'; this can occur when someone makes a new connection between ideas or sees a topic from a fresh perspective.

In this book, we are concerning ourselves with 'little-c' and 'mini-c' and realize that students may not be finding the key to unsolved problems (though you never know!). Rather, we are encouraging them to think differently about mathematics and to expand their notions of what *doing* mathematics entails.

Using the book

Our book is written with undergraduate students in mind and is suitable as a textbook for general education or teacher preparation mathematics courses. It offers a fresh approach to mathematics for general education courses – one that strays from the usual linear trajectory of the K16 curriculum. This approach may benefit preservice teachers as well by suggesting ideas for teaching mathematics for creativity in their future practice. The book could also be a reminder to prospective and practicing teachers of the beauty of teaching mathematics and the possibilities afforded by teaching for creativity – the reasons many of us teach and study mathematics in the first place.

That being said, the book can be used beyond the classroom. Examples include high school and college STEM clubs, summer bridge and enrichment programs, and homeschooling settings. The book is amenable to contemporary instructional platforms. It can be used to guide synchronous or asynchronous online courses. We also encourage recreational use of the book – put it on your coffee table and see what happens!

The quests

We begin the journey in Chapter 0 with an introduction before Chapter 1, which engages students in explorations that prompt them to consider the role of mathematics and creativity in their own lives. In Chapter 2, students are encouraged to make connections between mathematics and their surroundings and to think about how the introduction of certain constraints can lead to the need for creative solutions (Medeiros et al., 2018; Tromp, 2022; Tromp & Baer, 2022). Chapter 3 has students consider the role of play in mathematics and the role of mathematics in play: It includes explorations involving juggling, gummy bears, and finger paint. These types of explorations continue into Chapter 4, which focuses on the aesthetics of mathematics and the beauty around us that becomes visible through mathematics. In Chapter 5, we encourage students to wonder about their surroundings through their newfound appreciation for the creativity in mathematics. It is our hope that this sense of wonder will stay with the student long past their interactions with this book.

The quests are organized around the following guiding questions:

Chapter 1: What does it mean to you?
Chapter 2: What do you notice?
Chapter 3: Where is the mathematics in play?
Chapter 4: Where is the beauty in mathematics?
Chapter 5: What makes you wonder?

Each of these chapters consists of three mathematics-driven explorations, called quests, that relate to the chapter's unifying theme. We use the term *quest* deliberately; we hope that students will embrace the challenge of embarking on a search for connections between creativity, mathematics, and themselves. Although this book allows for self-guided explorations, if you are in a classroom setting, we recommend that students work on the same quest, with the instructor facilitating the explorations during class. Most quests are best completed by students working in small groups. Many activities involve the use of tools, both physical and virtual, for students to engage with mathematics and communicate their thinking. Others challenge students to create new representations of their thinking and their understanding of mathematics. We believe that students should be given the opportunity to experiment and play with mathematics; we have created quests that are intended to draw out the creativity we are seeking in our students.

We recommend having students complete the quests in the order suggested in the book. This will help contribute to their sense of journey. We do, however, expect that you will need to make choices about the tasks you have your students pursue. We acknowledge that the material presented in this book is more than can be explored in a single semester, especially if students embrace the invitation to see where the road takes them. Although each quest contributes to the journey, the quests are also discrete from one another: You can skip quests without losing much meaning. Furthermore, the 'Going beyond' and 'Let's reflect' sections were developed with the expectation that the instructor will select from among the prompts provided.

The projects

Each of Chapters 1–5 ends with a corresponding project idea that encourages students to personalize their learning with respect to the chapter's theme and explore an area of interest to them through the lens of creativity and mathematics. Each project is presented as a possible semester-long or final project: We leave it up to you to select the projects that make the most sense for your course. On the other hand, you might choose to adapt these projects and have students explore them at the end of each chapter. The length and depth of each project can be adjusted to the context and objectives of your course. We hope that the suggested projects provide a creative space for students to engage with their emerging ideas and serve as a springboard for them to connect creative thinking through mathematics with their interests and goals.

Teaching for creativity

The *Teaching for Creativity* section appears as Chapter 6. This chapter provides suggestions for implementing each of our quests. Following a short description of the objectives of the quest, we list the mathematics content explored, primary connections to creativity, suggested materials, pacing suggestions, tips for instructional decisions, and recommended further readings. Our pacing guide is based on 100-minute class meetings. We also offer commentary on how to engage students in the quests, including some examples of specific instructional moves that can foster creativity and elevate the level of student engagement and ownership of learning.

The Facilitator Guide

The *Facilitator Guide*, available through the link found in the Front Matter (www.routledge.com/9781032890791), accompanies this textbook. The guide supplements the textbook by providing worksheets with diagrams and tables in printable form and sample responses (written by mathematics graduate students) to all numbered questions in the quests. In keeping with the open-ended nature of the 'Going beyond' and 'Let's reflect' sections, we intentionally do not offer sample responses for those sections. Those prompts encourage students to be creative and personalize their learning for themselves. We hope you will be impressed by what they can create!

Cited and related works

American Association of Colleges and Universities. (2007). *College learning for the new global century: A report from the National Leadership Council for Liberal Education & America's Promise*. American Association of Colleges and Universities.

American Mathematical Association of Two-Year Colleges. (2018). *IMPACT: Improving mathematical prowess and college teaching*. Ch. 1–6. American Mathematical Association of Two-Year Colleges. https://www.amatyc.org/impact/

Beghetto, R. A., & Kaufman, J. C. (2007). Toward a broader conception of creativity: A case for "mini-c" creativity. *Psychology of Aesthetics, Creativity, and the Arts, 1*(2), 73. https://doi.org/10.1037/1931-3896.1.2.73

Chamberlin, S. A., & Moon, S. M. (2005). Model-eliciting activities as a tool to develop and identify creatively gifted mathematicians. *Journal of Secondary Gifted Education, 17*(1), 37–47. https://doi.org/10.4219/jsge-2005-393

Cilli-Turner, E., Satyam, V. R., Savić, M., Tang, G., El Turkey, H., & Karakok, G. (2023). Broadening views of mathematical creativity: Inclusion of the undergraduate student perspective. *Journal of Creativity, 33*(1), 100036. https://doi.org/10.1016/j.yjoc.2022.100036

El Turkey, H., Karakok, G., Tang, G., Regier, P., Savić, M., & Cilli-Turner, E. (2020). Tasks to foster mathematical creativity in Calculus I. In S. Karunakaran, Z. Reed & A. Higgins (Eds.), *Proceedings of the 23rd Annual Conference on Research in Undergraduate Mathematics Education* (pp. 588–597). RUME.

Ervynck, G. (1991). Mathematical creativity. In D. Tall (Ed.), *Advanced mathematical thinking* (pp. 42–53). Kluwer Academic Publishers. https://doi.org/10.1007/0-306-47203-1_3

Hanstedt, P. (2024). *General education essentials: A guide for college faculty* (2nd ed.). Taylor & Francis.

Kaufman, J. C., & Beghetto, R. A. (2009). Beyond big and little: The Four C model of creativity. *Review of General Psychology, 13*(1), 1–12. https://doi.org/10.1037/a0013688

Lithner, J. (2008). A research framework for creative and imitative reasoning. *Educational Studies in Mathematics, 67*, 255–276. https://doi.org/10.1007/s10649-007-9104-2

Medeiros, K. E., Steele, L. M., Watts, L. L., & Mumford, M. D. (2018). Timing is everything: Examining the role of constraints throughout the creative process. *Psychology of Aesthetics, Creativity, and the Arts, 12*(4), 471–488. https://doi.org/10.1037/aca0000148

Munakata, M., Vaidya, A., Monahan, C., & Krupa, E. (2021). Promoting creativity in general education mathematics courses. *PRIMUS – Problems, Resources, and Issues in Mathematics Undergraduate Studies, 31*(1), 37–55. https://doi.org/10.1080/10511970.2019.1629515

Regier, P., & Savić, M. (2020). How teaching to foster mathematical creativity may impact student self-efficacy for proving. *The Journal of Mathematical Behavior, 57*(2). https://doi.org/10.1016/j.jmathb.2019.100720

Runco, M. A. (2004). Creativity. *Annual Review of Psychology, 55*, 657–687. https://doi.org/10.1146/annurev.psych.55.090902.141502

Runco, M. A., & Jaeger, G. J. (2012). The standard definition of creativity. *Creativity Research Journal, 24*(1), 92–96. https://doi.org/10.1080/10400419.2012.650092

Savić, M., Satyam, V. R., Turkey, H. E., & Tang, G. (2022). Mathematical creativity at the tertiary level: A systematic review of the literature. In S. A. Chamberlin, P. Liljedahl & M. Savić (Eds.), *Mathematical creativity: A developmental perspective* (pp. 105–120). Springer. https://doi.org/10.1007/978-3-031-14474-5_8

Saxe, K., & Braddy, L. (2015). *A common vision for mathematics science programs in 2025*. Mathematical Association of America.

Sternberg, R. J. (2010, October 10). Teach creativity, not memorization. *Chronicle of Higher Education, 57*(8), 1–4. https://www.chronicle.com/article/teach-creativity-not-memorization/

Sternberg, R. J. (2015). Teaching for creativity: The sounds of silence. *Psychology of Aesthetics, Creativity, and the Arts, 9*(2), 115–117. https://doi.org/10.1037/aca0000007

Sternberg, R. J. (2019). Enhancing people's creativity. In J. C. Kaufman & R. J. Sternberg (Eds.), *The Cambridge handbook of creativity* (pp. 88–104). Cambridge University Press. https://doi.org/10.1017/9781316979839.007

Tromp, C. (2022). Integrated constraints in creativity: Foundations for a unifying model. *Review of General Psychology, 27*(1), 41–61. https://doi.org/10.1177/10892680211060027

Tromp, C., & Baer, J. (2022). Creativity from constraints: Theory and applications to education. *Thinking Skills and Creativity, 46*, 101184. https://doi.org/10.1016/J.TSC.2022.101184

Acknowledgments

This textbook is the result of many collaborations over the years. Just as teaching does not occur in isolation, neither does the writing of a textbook.

Ceire Monahan and Ashwin Vaidya developed many of these lessons with us as part of the leadership team of the original Creativity in Mathematics and Science project (NSF Award #1611876). We splattered paint, blew up beachballs, juggled, and threw beans at giant atlases as we developed, taught, reflected upon, and revised many of these lessons. Thank you for your part in making this book possible. To our students in MATH 106: Thank you for your willingness to explore with us!

Avigail Handel, our content editor, contributed both her mathematical expertise and her copy editing skills to this book, greatly enriching it. She also provided many of the example responses in the *Facilitator Guide*. Her keen eye for detail, thoughtful feedback, and tireless work over many months continually challenged us and drew us into the kinds of discussions we enjoy. Her enthusiasm for the spirit of the book and her calm encouragement sustained us throughout the process. We feel deeply fortunate to have her on our team.

Thank you to Preston Pietruszewski and Ashley Etheridge for creating many of our images and figures. Ashley also created all of the butterfly images at the beginning of the chapters.

We are grateful to Miloš Savić, Jessica Krim, Eileen Fernández, and Ceire Monahan for reviewing and providing written endorsements for the book. And to our other reviewers, Emily Olson and Erin Krupa: Thank you for your thoughtful feedback on earlier drafts. Special thanks go to John O'Meara, who not only reviewed the book but also wrote some of the sample student responses in the *Facilitator Guide*. Our reviewers' collective expertise on creativity, STEM education, and curriculum development was the basis of their thoughtful reactions to the book.

Thank you to Sean Gandini, our juggling collaborator, who created the videos in Chapter 3. He helped us make our course (and our textbook) interdisciplinary in unexpected ways. Thanks also goes out to Mac Adams, for his photo of his *Ophiuchus* sculpture; Deepak Bal, for his mathematical doodles; Melissa Spigelman, for the photo of the fish tanks; and Adam Leszczyński, for the model and photo of mathematical connections.

We are fortunate to work with a fantastic group of colleagues at Montclair State University. We'd like to acknowledge those we work most closely with on our STEM education projects: Nina Bailey, Joe DiNapoli, Eileen Fernández, Steven Greenstein, Nicole Panorkou, Ashwin Vaidya, Doug Larkin, Emily Klein, Monica Taylor, and Colette Killian. It is a joy to work with such knowledgeable, dedicated, and compassionate educators. A special thank you also goes out to Dr. Evan Maletsky, professor emeritus, whose teaching and scholarship on fractals guided the creation of Quest 4.3.

We are thankful for the support that Julia Dolinger and Sofia Cohen from Routledge Publishing provided us, from the moment we submitted our proposal through the process of refining it.

Personal thanks from Mika:

Gracias to my husband Carlos Molina, who brings joy to all aspects of my life. His encouragement throughout this process sustained me and made this work possible. Arigato to my parents, who encouraged my education and instilled in me a love for mathematics.

Personal thanks from Eliza:

To my husband, Damian, and our sons, Adam and Mark – thank you for your love, patience, and kindness. Your belief in me made this book possible! To my parents, Irena and Zbigniew, and my sister, Beata – thank you for always cheering me on. Your support held me on course every step of the way.

This material is based upon work supported by the National Science Foundation under Award No. DUE–1611876. Any opinions, findings and conclusions or recommendations expressed in this material are those of the authors and do not necessarily reflect the views of the National Science Foundation.

All unattributed images and photos were created by the authors.

0 Welcome to *Creative Thinking Through Mathematics*!

When you think about mathematics, what comes to mind? Do you remember a particular teacher? A particular lesson? Or, do you remember the mathematics topics you have learned over the years? When we ask our undergraduate students these same questions, we often hear phrases such as 'PEMDAS', 'formulas', 'equations', and 'numbers'. Others talk about processes such as 'multiplication', 'solving for *x*', and 'graphing'. What does mathematics mean to *you*?

What about creativity? What comes to mind when you think about creativity? Perhaps a sculpture or a poem? A dance movement? Can you think of examples outside of the arts? When was the last time you were creative? What was creative about what you did? Have you ever thought about mathematics as being a creative discipline? We wouldn't be surprised if you have not. In fact, most students we have surveyed consider mathematics to be not very creative at all (Munakata & Vaidya, 2012). We are here to challenge the notion that when you 'do' mathematics, you only work with formulas, definitions, procedures, and calculations. Mathematics can be so much more: In fact, mathematics can be creative! In this book, we will ask you to tap into your creative self to think deeply about mathematics.

At this point, it would be fair for you to ask us about *our* notions about creativity and mathematics. We are motivated by various definitions of creativity that say that it requires both novelty and usefulness (e.g., Runco & Jaeger, 2012). But what do we mean by *novelty*? When something is new, unexpected, or different, it might qualify as being novel. In the context of this course, novelty might be related to an idea or interpretation of a problem. It might also take on the form of a novel approach or solution to a problem. Your idea might be novel because it is different from your classmates' ideas or because it is an idea that is new specifically to you.

In this book, you will be asked to consider mathematical problems in unexpected or new ways, and just as importantly, you will be asked to argue how your specific approach and solution make sense mathematically (and thus be *useful* to the discussion). Our considerations

DOI: 10.4324/9781003541097-1

for the *usefulness* of an idea go beyond direct applications of mathematics to 'real-world' problems: Ideas are useful if they are relevant to the problem and shed light on the phenomenon being explored. This book encourages you to think in novel ways and contribute useful ideas to the conversation.

Another notion about creativity that motivates us is that it can be learned and cultivated (Sternberg, 2019). We reject the notion that only some of us are born to be 'creative'. Humans are creative beings. We all engage in creative thinking in our everyday lives when facing life's struggles, finding alternative ways to overcome obstacles, or reaching our personal and professional goals (Richards, 2010). We are creative when we take it upon ourselves to use what we know to make something new or to improve something. We are creative when finding a solution after our shoelace breaks; we are creative when decorating birthday cakes; we are creative when corralling rattlesnakes (which we hope is not an everyday situation!) Just as we can be creative in our lives more generally, we can be creative when making sense of a mathematical situation. This book connects different ways of being creative with mathematical thinking: You will enhance your creative thinking as you explore mathematics in a unique way.

Finally, not only do we believe that everyone can be creative, but we also believe that everyone can do mathematics. What we said earlier about being a creative person, we can say again about being a math person: We reject the notion that only some people are born to be 'math people'. Everyone is a 'math person'. In fact, whether we are aware of it or not, all of us are constantly engaging in mathematical thinking. Sometimes, it is in day-to-day situations, like planning a series of errands or determining shift schedules for employees at a restaurant. At other times, it might be in class, when interpreting a graph or visualizing the rotation of a figure. Contrary to popular belief, mathematical thinking does not even have to involve numbers and computations and often involves creative approaches and solutions. We can all act in new or unexpected ways when engaging in mathematical activities, if given a chance. We believe that mathematics promotes creativity and that through creativity, everyone can personalize their learning of mathematics. In turn, by taking ownership of our learning of mathematics, we begin to make new meanings, notice new connections, and wonder about the world around us through a mathematical lens.

As you may have guessed by now, we believe that mathematics and creativity belong together. We leverage this special connection to challenge you to engage in *creative thinking through mathematics*. Throughout the book, you will see that mathematics is front and center. It will both spark your creativity and anchor your creative thinking. Regardless of what your last mathematics class was, or if it has been a while since you felt like you were intentionally doing mathematics, this book will take you on a journey that engages you with mathematics in a way that may be new to you. We hope that the book opens your eyes to what is possible when you use creative thinking to make sense of mathematics.

How to use the book

This book starts with inviting you to think about what place mathematics and creativity have in your life. It will then ask you to pay close attention to the world around you as you look through a fresh lens and consider the intersections of mathematics and creativity. By the end of the journey, you will have delved into deep mathematical problems, some of which have yet to be solved (by anyone)! We anticipate that you will come to see yourself in a position to continue wondering and asking questions about the world around you through mathematics and creativity.

This book contains seven chapters, numbered 0 through 6. While all of these chapters serve important purposes – Chapter 0 is an introduction to students and Chapter 6 is our *Teaching for Creativity* chapter – Chapters 1 through 5 are the true heart of the book. Each of these five chapters is centered around a theme tied together by a guiding question:

Chapter 1: What does it mean to you?
Chapter 2: What do you notice?
Chapter 3: Where is the mathematics in play?
Chapter 4: Where is the beauty in mathematics?
Chapter 5: What makes you wonder?

Within each of these chapters, you will find three quests. The word 'quest' illustrates our intention to offer each set of activities as part of a personal journey in which you consider creative approaches to mathematics. You may think of each quest as an opportunity to search for deeper meaning, not only in a mathematical sense but in other disciplines as well. The quests engage you in conversation through questions posed throughout. Each quest ends with 'Going beyond' and 'Let's reflect' sections that challenge you to make this all mean something to you and encourage you to tap into your creative self for ideas, questions, and personal connections. Your instructor may choose the questions and activities for the class to focus on, but we also invite you to take your own deep dive into these sections and complete as many tasks as you can: We pose many tasks, so choose the ones that inspire you to be creative in bringing new ideas to life.

Each quest follows roughly the same format:

I. *Invitation.* We start by posing situations to ponder related to the mathematics in the quest.
II. *What do you notice?* This question invites you to think of experiences or situations related to the quest. The heading for this section will vary from quest to quest.
III. *Let's see what happens.* This is an invitation to play and explore mathematics through creativity.
IV. *What else?* We ask you to acknowledge your emerging ideas as you continue to develop them.
V. *Story.* We provide you with a glimpse into the context or history relevant to the quest.
VI. *Going beyond.* We pose problems and questions that encourage you to follow your interests and use your skills to pursue new avenues that merge mathematics and creativity.
VII. *Let's reflect.* Reflection is a critical part of the process of learning. Here, we suggest activities that can guide your reflections.

Throughout the book, you will be asked to doodle, juggle, stroll, and construct as you wonder about the mathematics around you. We encourage you to take ownership of your own learning. Resist the temptation to just 'Google' the answers and be mindful of how you use artificial intelligence (AI). It may make sense to go online to review mathematics terminology or notation or find specific information needed for a particular quest. Or, perhaps you feel the need to use online resources to review and complete practice problems related to some of the quests. Those are all great uses of technology. For the most part, however, looking up answers to questions we pose beyond these particular uses will not support your creativity. In fact, it will undoubtedly curtail it. Take up the challenge, don't be afraid to not 'get it' right away, and embrace the struggle. It is possible that you will feel uncomfortable at times not knowing how to proceed, but that is absolutely okay. If you feel uncertain about the next step to take, it just means that you have an opportunity to be creative.

Whether you are a student, an instructor, or a lifelong learner (or all three!), we hope that this book encourages you to tap into your creativity as you explore mathematics. Creative thinking can support imaginative explorations in mathematics, while mathematical reasoning serves as a framework for enhancing the creative process. We hope you embrace this opportunity to challenge yourself in ways that expand your views about both mathematics and creativity. So, gather your curiosity, inquisitiveness, and imagination, and let's start exploring!

A 'creativi-tree' developed by former students is presented in Figure 0.1. Each branch represents a characteristic of creativity, with the leaves representing connections between mathematics and creative thinking. As you explore the quests in this book, you will be asked to jot down your own notions of how you are being creative in mathematics. We encourage you to record examples of special actions or ideas that you were proud of in each quest. You can then make your own 'tree' or another design that reflects your personal journey through the landscape of mathematics and creativity!

Figure 0.1 A 'creativi-tree' of mathematics and creative thinking characteristics created by former students.

Questions to reflect on before moving on to Chapter 1:

1) What is my approach to creativity?
2) What is my approach to mathematics?
3) What are my expectations for this book?

Cited and related works

Munakata, M., & Vaidya, A. (2012). Encouraging creativity in mathematics and science through photography. *Teaching Mathematics and Its Applications: An International Journal of the IMA, 31*(3), 121–132. https://doi.org/10.1093/teamat/hrr022

Richards, R. (2010). Everyday creativity. In J. C. Kaufman & R. J. Sternberg (Eds.), *The Cambridge handbook of creativity* (pp. 189–215). Cambridge University Press. https://doi.org/10.1017/CBO9780511763205.013

Runco, M. A., & Jaeger, G. J. (2012). The standard definition of creativity. *Creativity Research Journal, 24*(1), 92–96. https://doi.org/10.1080/10400419.2012.650092

Sternberg, R. J. (2019). Enhancing people's creativity. In J. C. Kaufman & R. J. Sternberg (Eds.), *The Cambridge handbook of creativity* (pp. 88–104). Cambridge University Press. https://doi.org/10.1017/9781316979839.007

1 What does it mean to you?

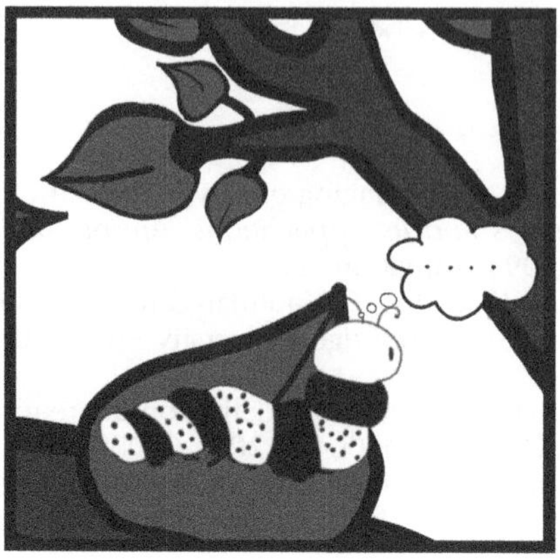

What does creative thinking through mathematics mean to you?

Through the three quests in this chapter, you will begin to see that mathematics goes beyond just numbers and computations. Often, mathematics problems have no one, clear answer. Maybe you have faced situations (in mathematics or other contexts) where ambiguity made you uncomfortable. We hope that you will come to embrace a little bit of messiness. Many mathematical tasks can be open to multiple perspectives, approaches, and solutions. We encourage you to interpret problems in your own way, test out different approaches, and consider alternative solutions. By encouraging imagination (Ervynck, 1991), flexibility (Torrance, 1965), and risk-taking (Beghetto et al., 2021), these explorations will not only tap into your creativity but also make the process of learning personal to you (Munakata & Vaidya, 2013).

This chapter aims to make mathematics resonate with you in other ways as well. You will consider how mathematics and creativity relate to your current situation and to your academic, career, and personal goals. You will also explore how mathematics can be used to provoke curiosity, discussion, and action around global, societal, or community issues of importance to you.

Chapter 1: What does it mean to you?

DOI: 10.4324/9781003541097-2

Quest 1.1: Examining our relationship with creativity and mathematics

Let's connect the dots!

This first quest in our journey has you thinking about the role of creativity and mathematics in your life. You will be asked to consider tasks from multiple perspectives and to begin thinking about the connections between mathematics, creativity, and yourself. By starting in this way, we hope to ground the explorations in this book to your ways of thinking, exploring, and wondering so that the learning of mathematics through creativity becomes a personal experience for you. Let's get started!

I. Invitation

Connect the two dots in Figure 1.1:

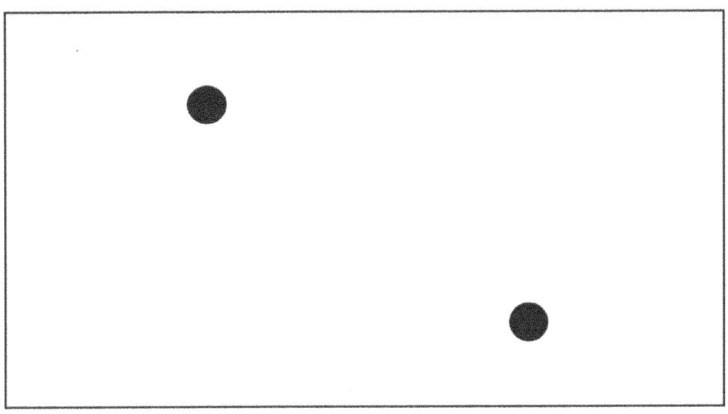

Figure 1.1 Two dots for connecting.

1) How did you connect the dots? Describe how you connected them.
2) Did you connect them like this (see Figure 1.2)?
3) Can you think of another way to connect them? Try it!

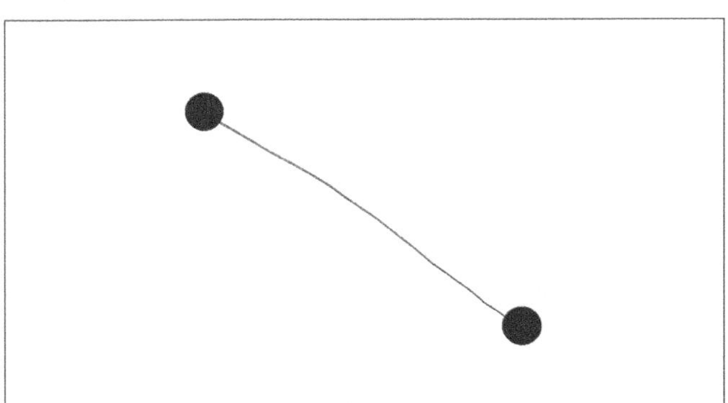

Figure 1.2 Two dots connected with a hand-drawn straight line segment.

We recently gave this task to 123 undergraduate mathematics students. Of the 123 students, 94 (roughly 76%) drew a line segment connecting the two dots in the first attempt, similarly to what is shown in Figure 1.2. Does this surprise you?

Try to connect the dots in three different ways.

Take a look at this example of a student's initial response (Figure 1.3):

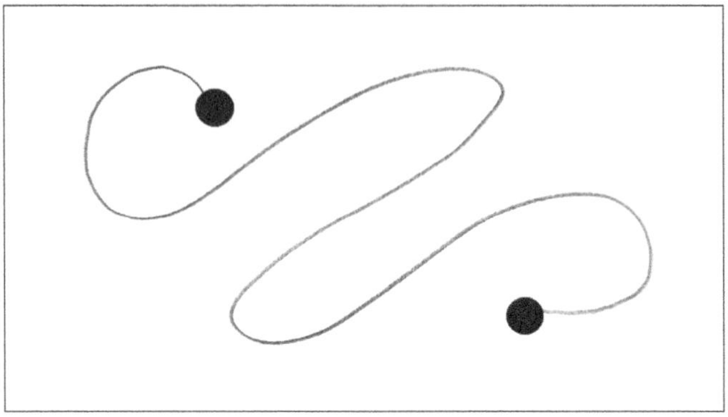

Figure 1.3 An alternative way to connect the two dots.

Look back at your attempts at connecting the dots.

4) Which was the most creative 'connection' you made?
5) How else might one be creative in connecting two dots?

For this activity, we asked you to doodle. A doodle is a casual sketch or drawing that is meant to help you think through a problem or represent an idea. Believe it or not, doodling is a very useful way of engaging with mathematics. We recently took a photo of a chalkboard used by mathematics researchers (both faculty and students) at our university (see Figure 1.4). You can see how they used doodling to try out new ideas and communicate mathematics.

Most of us would probably assume at the outset that we were being asked to connect the two dots with a straight line segment, which is also the shortest path between the dots. This approach shows a focus on precision and efficiency, which is valid and useful and has its place in mathematics. In this activity, however, you were asked to focus on more creative approaches, which could involve interpreting the instructions differently, questioning your assumptions, and determining alternative interpretations and solutions. This kind of thinking is useful in mathematics, where innovation often results from questioning, reinterpreting, and conjecturing. By doodling alternative ways to connect the two dots (a seemingly easy task), you were being creative and engaged in divergent thinking. Divergent thinking, often associated with creativity, describes one's ability to think of many possible alternative approaches, uses, and solutions (Guilford, 1950). We saw that when interpreted differently, the same task can lead to myriad other possible solutions.

In Figure 1.5, we list some of the features of creative thinking that have been proposed by researchers. Some refer to the process of being creative, while others describe the idea or

Figure 1.4 Doodles by Dr. Deepak Bal, Montclair State University mathematics professor, who had no idea his work would end up in this book (but who later gave us permission to use this photo!)

product. Throughout the book, you will be challenged in some, if not all, of these ways. We also expect that by the end of this journey, you will have added your own characteristics of creativity to this list.

Here are some features of creativity you will be asked to embrace in your approaches to mathematics:

- **Be open to new ideas** (Feist et al., 2019; Oleynick et al., 2017). Be curious and motivate yourself to explore.
- **Think of ideas that are original and useful** (Runco & Jaeger, 2012). The ideas can be original just to you and useful to the mathematics being discussed.
- **Strive for fluency** (Guilford, 1950; Leikin & Lev, 2013). Try to come up with many ideas related to a problem.
- **Be flexible** (Torrance, 1966). How can you get unstuck? How do you overcome obstacles?
- **Seek personally meaningful insights** (Kaufman & Beghetto, 2009). What new perspectives have special meaning for you?
- **Provide elaboration** (Lev-Zamir & Leikin, 2011). How can you explain your reasoning to others?
- **Be imaginative** (Csikszentmihalyi, 1997; Gotlieb et al., 2019). What approaches, paths, or scenarios can you envision?
- **Be comfortable with ambiguity** (Zenasni et al., 2008). Try to embrace uncertainty as inspiration to be creative.
- **Take risks** (Mann, 2006; Sternberg & Lubart, 1995). Have an idea? Just try it and see what happens!

Figure 1.5 Some features of creative thinking, with connections made to mathematics, including literature citations.

II. What is creativity to you?

Think of times when you were challenged to be creative, either in your day-to-day activities or in a mathematics class. Copy and complete Table 1.1, filling each cell with an example of an activity where you engaged with creativity in that setting. Add other traits of creativity that capture some of your experiences in the last three rows and give examples of each.

Table 1.1 Blank table to be copied and filled in with examples of creativity in day-to-day and mathematical activities.

	Day-to-day activities	*Mathematics class*
Being open to new ideas		
Taking risks		
Tolerating ambiguity		
Being flexible		
Being imaginative		

1) What other words come to mind when you think of creativity?
2) How is creativity useful in your academic pursuits?
3) How is creativity useful to your other goals?
4) What tasks or activities make you feel creative? Why?
5) What (if anything) has hindered your ability to be creative in the past? Have you only felt creative in certain settings?

III. Let's see what happens

Figure 1.6 shows a representation of a round cake. Can you cut a round cake into eight pieces with three straight cuts using a standard knife? How?

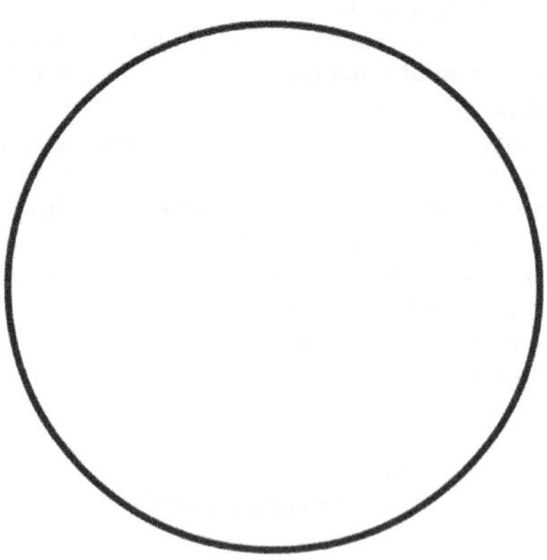

Figure 1.6 A circle representing a round cake.

1) Were you able to cut the cake into eight pieces? If not, how close did you get?
2) What obstacles did you face?
3) How did you try to overcome these obstacles?

Now, imagine yourself sitting at a table with a real cake.

4) Does this mental image help?
5) What if we gave you additional tools?

Get hold of an actual cake or make your own cake with Play-Doh®.

6) Does having a cake (whether an actual edible cake or your Play-Doh® cake) at your disposal help?

It's okay to continue wondering about whether or not it is possible to cut a cake in this way. Creativity is not something we can call upon in an instant. It takes time and sometimes inspiration will hit you at the most unexpected moment. It might be at the grocery store when you notice a colorful birthday cake. It might be on a walk, as you are appreciating the falling leaves. Often, when we are stuck, the best action to take is to walk away and come back to the problem. So, if you're stuck, go for a hike, think about the clues, and let that 'aha' moment come to you!

IV. What else?

Let's keep going with the cake example. Assume that you are restricted to cutting with a standard knife and that each cut goes through the entire cake (as opposed to partway through). Consider the smallest and largest number of pieces that result from making cuts to a cake that is 1) represented as a circle on paper and 2) represented in three dimensions using Play-Doh®. Copy Table 1.2 and record your findings for the numbers of cuts shown.

Table 1.2 A table for recording the number of pieces that result from cuts through a round cake.

Number of cuts	*Circle drawn on paper (2-dimensional)*		*Play-Doh® cake (3-dimensional)*	
	Smallest number of pieces	*Largest number of pieces*	*Smallest number of pieces*	*Largest number of pieces*
1				
2				
3				
4				
5				

1) What do you notice?
2) What are you wondering?

V. Story

Believe it or not, cake cutting has been studied extensively in mathematics. Not only does cake cutting have implications for how much or what piece of the cake you might get, but it is also used as a model for determining the most 'fair' way of divvying up something of value (like land) among entities (like people). This has important implications for issues at even the global level, for example, when countries negotiate the distribution of resources.

So how is cutting a cake mathematical? By wondering, pondering, struggling through the problem, and arriving at a creative solution, you were engaging in mathematical thinking. Mathematics is not always about the solution; it is often about the *process* of finding a solution or solutions. The solution to any worthwhile problem should not come to you automatically – that would not help you grow. In fact, a problem for which an answer is immediately apparent is just an exercise (Schoenfeld, 1985). It is the process and the learning that takes place that is creative and mathematical.

Through your experiences, interests, and goals, you have undoubtedly engaged in creative thinking in multiple ways. Creativity is not necessarily something you are born with (Sternberg, 2019). Rather, it is a way of thinking that you can cultivate through experiences and your interactions with others.

Throughout this book, you will have the opportunity to personalize your learning of mathematics through creativity. You will build on your prior knowledge and experiences and engage with creativity and mathematics in a way that is personal to you.

VI. Going beyond

1) Creativity is very difficult to define. So far, we have given a few definitions in this text. How would *you* define it?
2) Think of words associated with creativity and with mathematics. Consider words associated with creativity to be in one set (collection of items) and words associated with mathematics in another. Create a Venn diagram, placing each word in the appropriate set (or section of the diagram). Your Venn diagram could take the form of one of the diagrams in Figure 1.7. Modify the diagrams as needed to describe your thinking. Justify your placement of each word.
3) Now think of words related to your academic major or career interests. Add these words to your Venn diagram from Question 2. Describe what your Venn diagram reveals.
4) Give the 'connect the dots' task to your friends and family (or to strangers on the street) and see what they do. Keep track of the results and present them in a creative way.
5) Create a task similar to the 'connect the dots' activity that is open to multiple perspectives. Be sure to make your task such that it invites different responses. What elements of the 'connect the dots' task are you incorporating? Test it out on friends, acquaintances,

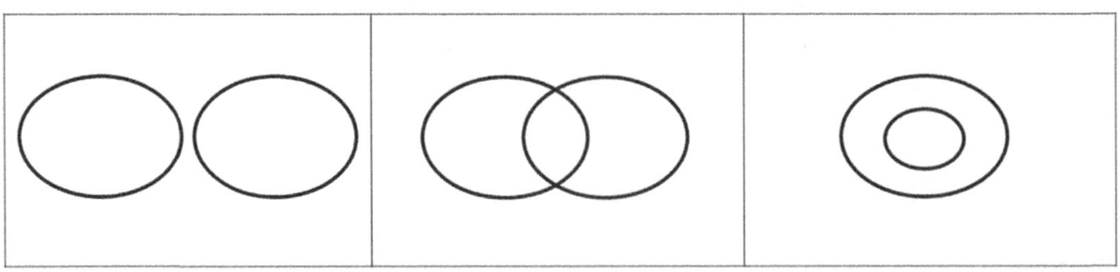

Figure 1.7 Examples of three types of Venn diagrams: disjoint sets (left), overlapping sets (middle), and nested sets (right) to use for Question 2.

and strangers. What were their different responses? Which responses do you think were the most creative?

6) What other mathematics problems have you encountered that called on you to be creative?
7) List people who you think are creative and consider what they have in common. Think of different categories of people including

    a) Famous people
    b) Friends and family
    c) People from your community

8) We mentioned that cake cutting can be used to model how to allocate items fairly. Look up the mathematical ideas behind fair division and write about what you learn. For example, how many different ways can you find to divide property (or cake) between two people? Five people?

VII. Let's reflect

1) When did you last feel creative outside of mathematics? Describe the experience.
2) When did you last feel creative in a mathematics classroom? Describe the experience.
3) Keep a creativity journal for a week. Track all of the different ways in which you were creative in your everyday life. Which ones surprised you?
4) What is the most creative idea or object that you have ever encountered? In what ways did it exemplify creativity?
5) Recall a mathematical idea or solution that made an impression on you. What made this idea so special to you?
6) Do you think that mathematics is a creative discipline? Why or why not? Give specific examples.

Quest 1.2: Creative approaches to reasonable estimates

Did someone order a magnitude?

In mathematics, we often spend our time trying to find the 'right' answer and trying to be as exact or precise as we can. But when might being precise not be the goal? In fact, when might it not be possible to know an exact answer? In this quest, we will explore different situations when a guess is 'good enough'. We will see how we can tap into our creativity to find a reasonable estimate in these situations.

I. Invitation

In the previous quest, you were asked to view situations from different perspectives. You saw that there were multiple ways to connect two dots. You may have struggled with cutting a cake until you saw it from a different perspective. Perhaps having access to different tools (like Play-Doh® or an actual cake) helped shift your perspective in a useful way. Now you are going to consider a different aspect of mathematics and draw from your experiences to determine what is 'reasonable'.

You will be asked to develop strategies to estimate a value or quantity within a reasonable range. You will think about different situations when an exact answer isn't necessary, appropriate, or even possible. In this quest, you will interpret and make sense of ambiguous situations and find innovative ways to develop and justify your estimates. One criterion we can use to consider the reasonableness of answers is called **order of magnitude,** which we define later. You will then determine whether or not your estimates fall within an order of magnitude of the actual value (if one exists). The somewhat 'random' questions we pose will give you practice in posing and solving problems that you can apply in contexts that have more meaning for you in the future.

II. What does it mean to you?

Recall what you have done or experienced in the last 24 hours. We bet that you not only engaged in mathematics but called on creativity as well. Believe it or not, our minds are constantly 'doing' mathematics and being creative even when we are completing mundane tasks. What is the most efficient path to walk to the store? How do I fit all of these books into my bag? How many onions should I buy? The answers to these questions involve both mathematics and creativity. This quest asks you to draw on your mathematical knowledge in a creative way to determine estimates for situations when a guess is 'good enough'. Note your responses to the following prompts:

1) Can you think of situations when it wasn't worth the effort or time to determine an exact answer?
2) When in our everyday lives is an estimate 'good enough'? When is it, in fact, preferred?
3) Are there situations when an estimate is the best you can do?
4) What professions rely on estimates?
5) What determines the reasonableness of an estimate? What makes it 'good enough'?

III. Let's see what happens

This quest calls on you to develop an estimate that is 'good enough'. When we measure or estimate a value, we can aim to be precise or accurate, or both. Precision has to do with being exact or consistent in your measurements. On the other hand, **accuracy** has to do with how close to the actual value you are.

One can determine whether or not an estimate is satisfactory by checking if it falls within an order of magnitude of the actual value, assuming that the actual value can be determined. To be within an order of magnitude of a number, an estimate must be within a factor of 10 of the actual value. That means that it must fall within a range between $\frac{1}{10}$ of the actual number and 10 times the actual number. So, for example, a number within an order of magnitude of 58.27 would be between 5.827 and 582.7. We would consider 193 to be within an order of magnitude of 58.27 because it falls between 5.827 and 582.7. On the other hand, 3.2 would fall outside of this range, as would 2,109.41.

1) What is the range of values that are within an order of magnitude of 13.22?
2) Is 240,214 within an order of magnitude of 3 million?
3) Give two numbers that are within an order of magnitude of 0.00718, one smaller than and one larger than 0.00718.
4) Give two numbers that are *not* within an order of magnitude of 0.00718, one less than the lower bound and one greater than the upper bound.

IV. What else?

In this section, we present some order of magnitude questions. First, note how you interpret each question. What assumptions are you making? What parameters or restrictions will you consider?

Once you have made sense of the question, determine an estimate that you think is within an order of magnitude of the actual value. Be ready to describe your process and defend your answer. Keep notes on your initial strategy and how it evolves as you consider each question in turn. In what ways are you being creative as you form an estimate? What experiences are you drawing on? Scribble, doodle, and think out loud, but resist the temptation to use technology beyond a basic four-function calculator. Remember to include the units where applicable – they are very important. (Imagine: Knowing that an object's length was measured in centimeters instead of kilometers tells you more about the object than knowing only the numerical value of the measure.)

Consider the following order of magnitude questions:

1) What area of the Earth's surface does the Pacific Ocean cover?
2) What is the thickness of a piece of paper?
3) How much does a human eyelash weigh?
4) How many steps do you take in a year?
5) How tall is an ant?
6) How many trees are there in your neighborhood?

7) What is the total length of all of the rivers in your geographic area?
8) What is the length of the axon of a nerve cell?
9) What is the weight of a goldfish tooth?
10) How much does a cumulus cloud weigh?

Let's reflect on how you used creativity to answer the questions. Respond to the following for each of the problems:

1) In what ways were you creative in answering the question?
2) What assumptions did you make in your interpretation of the question?
3) What was your initial strategy for that question?
4) What elements of creativity did you draw upon as you came to your estimate?
5) What experiences did you draw upon?
6) Describe your process and your thinking, step by step.
7) If you are confident that your answer is within an order of magnitude of the actual value, how would you convince someone of that?
8) Which question required the most creativity to answer, and why?

V. Story

Enrico Fermi (1901–1954) was an Italian and American physicist who is best known for his pioneering work in nuclear physics and his role in the Manhattan Project, the American top-secret research mission that led to the production of atomic bombs during World War II. As a lecturer, he would famously ask his class how many piano tuners there were in Chicago. Although seemingly nonsensical – when would anyone ever need to know that? – this question calls on us to be creative in our approach. The parameters are ambiguous: Should we include the suburbs of Chicago? Do retired piano tuners count? Because the parameters of this question can be interpreted in different ways, we are called upon to set our own parameters and be creative as we consider different perspectives to answer it. This is one example of a whole set of questions that require some creativity and thoughtfulness before even attempting to formulate an answer. The answers to these questions must be bound by mathematics because the estimates should make sense in the context, be reasonable, and be justifiable. What's interesting to us about the solutions isn't the actual estimate you propose but how you get there and whether or not the answer can be justified.

These types of questions are also referred to as 'back of the envelope' problems. They're the type of problems that you can attempt to find an answer to by doing some quick calculations on the back of an envelope. Of course, with today's technology, we have the capacity to search for and find answers to many of these questions in an instant, but what fun is there in just knowing the answer? With just your mind, a pen and paper and (perhaps) a simple calculator, you can exercise your mental flexibility and use mathematics to reason about the world.

VI. Going beyond

1) Consider your academic and non-academic interests. Come up with five order of magnitude questions related to each of these.
2) Look for patterns in your solutions to the problems in this quest. Was there a particular strategy you utilized more than others?

   a) Choose one problem you solved in your usual way.

b) Now, try to solve it in a different way, using what you consider to be a completely different strategy.

3) Pose three order of magnitude problems that would be impossible to answer.
4) Pose three order of magnitude problems that suggest very small numbers.
5) Make up an order of magnitude problem. Take an informal survey: Ask your friends and neighbors (in person or through social media) to respond with their estimates. Analyze those estimates. How many fell within an order of magnitude of the actual answer or of the value you had in mind?
6) Review the definition of *order of magnitude*. This is just one parameter we can use to determine what is reasonable. In what situations is using order of magnitude to determine the reasonableness of estimates useful? When is it not useful? In what way could using the order of magnitude be harmful or detrimental? What other methods are there to determine the reasonableness of estimates?
7) Recall scientific notation. If you need a refresher, use technology to do some research. How are order of magnitude and scientific notation connected?
8) Recall or look up logarithmic scales. How are order of magnitude and logarithmic scales connected? What are the advantages and disadvantages of using the logarithmic scale to represent numbers?
9) Scientists often use estimation in their work with very large quantities, like the number of stars in our galaxy or the population of fish in a lake. Choose a topic and research how scientists typically make estimates related to that topic.

VII. Let's reflect

1) Keep an estimation journal for a day. Note all instances during the day when you catch yourself estimating. What patterns do you see? Is there a pattern to the types of situations you encounter where an estimate is 'good enough'? Is there a pattern to how you approach these tasks?
2) Reflect about the role of estimation in your life. How do you engage in creativity when you estimate?
3) What careers require one to estimate?
4) What role will estimation play in your intended career?
5) When is estimation enough? When is it not enough?
6) What methods help us improve our estimates?
7) Have you ever estimated something inaccurately? Why did it happen? What were the consequences? What could you have done differently?
8) Can we ever be completely precise? If yes, give an example of the object or situation and how we can know that the measurement is precise. What could someone say to argue that it is not precise?

Quest 1.3: Designing infographics to represent what matters

Communication matters

What are some issues or causes that are close to your heart? What do you care about? Are your concerns global? Societal? Local? Personal? How might you convey data related to what matters to you in an effective and appealing way?

One way to communicate information in a visually appealing, succinct way is through **infographics**. Infographics are graphic designs meant to combine mathematics and art by displaying data accurately through pictures or artistic diagrams. Infographics are often presented within the context of the relevant issue, and their purpose is, in most cases, to share results of data analysis in a way that can be understood by the general public. Magazines, newspapers, and TV ads often show numerical data, survey results, and other research findings through infographics. In this quest, you will analyze infographics and create your own infographic for a cause that is important to you. This will be your chance to pursue issues close to your heart in an artistic, personal, and mathematically accurate way.

I. Invitation

In the previous quest, you were creative in how you found reasonable estimates for quantities and measurements that were otherwise difficult or impossible to determine exactly. You tried to be as accurate as you could, within an order of magnitude. Here, we focus more closely on how to represent data accurately. You are going to use your creativity to represent data so that others can quickly and easily access information about an issue that is important to you.

You are probably aware that in mathematics, numerical data is often displayed in tables and graphs to make it easier to understand. What you might not know is that infographics that go beyond the simple table or graph – and are meant to be more appealing – also rely heavily on mathematics.

II. What issues are important to you?

We live in a complicated and ever-changing world. Each of us is affected by various societal, environmental, or policy matters. What problems came to mind as you read the introduction to this quest? How could you personally contribute to making the world a better place? Perhaps you are passionate about the environment or equitable access to housing. Maybe there is a problem in your community that needs a solution. There could even be a situation in your household that you would like to address.

Consider the infographic in Figure 1.8.

1) What information does this infographic communicate? What is the main message of this infographic?
2) How is this infographic different from a typical graph?
3) What does it make you wonder about?
4) In what ways does the infographic use mathematics to communicate data?
5) In what ways does the infographic exemplify creativity?
6) What are some limitations of the infographic? What information would tell a more complete story or help you better understand the phenomenon?

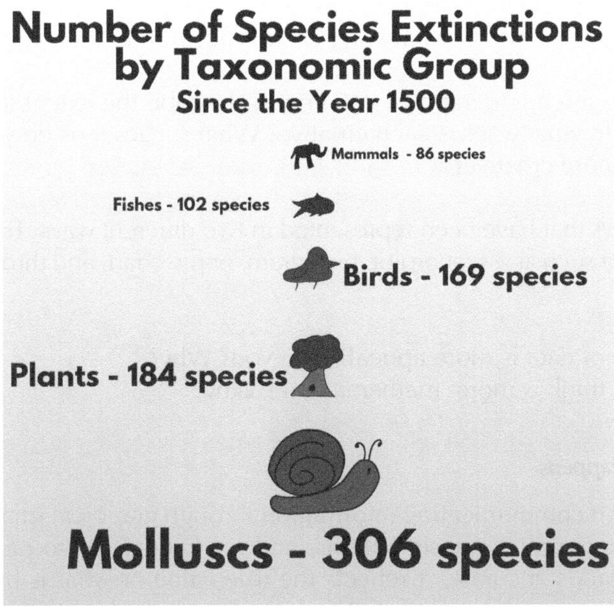

Figure 1.8 An infographic of species extinctions by taxonomic group since 1500, based on a bar graph retrieved from https://ourworldindata.org/grapher/extinct-species-since–1500. Infographic by Ashley Etheridge.

We can consider Figure 1.8 to be an infographic because it displays data through images that are both easy to understand and visually appealing. Notice how few numbers are displayed and that the size of each representation of the five different taxonomic groups is roughly proportional to the number of extinctions of species within that category. Although the term infographics can be used to describe many visual displays of data (or messages more generally), we will use this term specifically to refer to displays of data that go beyond the traditional forms (for example, histograms, scatterplots, and pie charts). The infographics in this quest call on you to notice how it is possible to use creativity to communicate data in a mathematically sound and visually appealing way.

Use the Internet to find some infographics on matters of importance to you. Make sure that each depicts a mathematical representation of data. Find three that you'd like to focus on. Be sure to cite the source of each graphic.

Consider the following questions:

7) Why did you choose these three infographics to analyze?
8) Based on the three infographics you found, what would you say are some advantages of using infographics instead of a more typical graph (for example, scatterplot, histogram) or chart?
9) What are some drawbacks of using infographics?
10) What mathematical content and techniques does each of your chosen infographics showcase? In other words, where is the mathematics in each infographic?

11) For each of the three infographics you chose, describe the extent to which each is mathematically accurate. In what ways is each accurate? How could their accuracy be improved?
12) For each of the three infographics you chose, describe the extent to which it exemplifies creativity. In what ways is each creative? What suggestions do you have for making each of them more creative?

Find data in the news that have been represented in two different ways: Through a more traditional display of data such as a scatterplot, histogram, or pie chart, and through a more creative infographic.

13) Which display of data is more appealing to you? Why?
14) Which do you think is more 'mathematical'? Why?

III. Let's see what happens

Being accurate when communicating information through graphical images is important for consumer confidence, policy considerations, and marketing, just to name a few contexts. When an infographic is accurate, it reflects the true value of what is being measured and represented. This requires the use of mathematical tools and methods, including scaling (proportions), graphing techniques, and proper representations of data.

Figure 1.9 shows a pie chart of the relative amount of municipal solid waste (MSW) that the United States population recycled in 2018. Pie charts are often used to compare percentages. Although not visually ornate, pie charts can be considered to be infographics because they transform raw data into visual formats.

The same data used to create the pie chart in Figure 1.9 has been used to create the infographics in Figure 1.10 and Figure 1.11.

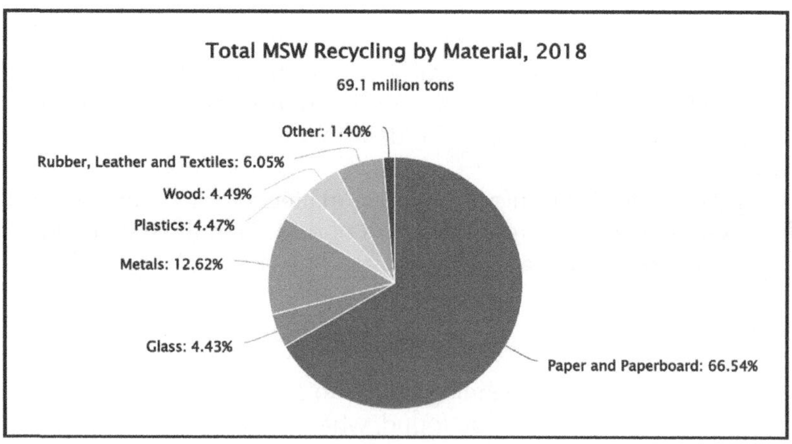

Figure 1.9 A pie chart of total municipal solid waste (MSW) recycling by material in 2018, retrieved from the United States Environmental Protection Agency (EPA) at https://www.epa.gov/facts-and-figures-about-materials-waste-and-recycling/national-overview-facts-and-figures materials

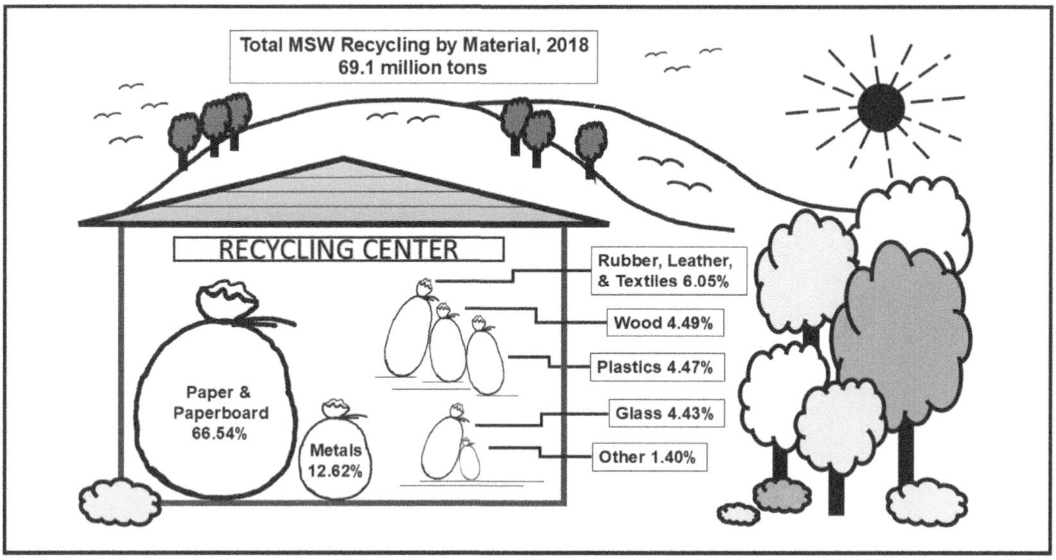

Figure 1.10 An infographic of the EPA's MSW recycling by material data in 2018, based on the pie chart published by https://www.epa.gov/facts-and-figures-about-materials-waste-and-recycling/national-overview-facts-and-figures-materials

Study the infographic in Figure 1.10 and answer these questions:

1) What are some key points of the infographic?
2) What are some advantages and disadvantages of representing data in this way?
3) Which do you prefer – the pie chart (a more traditional infographic) in Figure 1.9 or the infographic in Figure 1.10? Explain your answer. Can you see how others might prefer the other representation? What might their response be?

A good infographic makes it easy to read its content: It is informative, accurate, and aesthetically pleasing. But have you ever been misled by an infographic? Whether intentionally or not, graphs can mislead us, especially when they have missing data, disregard proportional scaling, use visual perspectives in a misleading way, or mislabel attributes.

Figure 1.11 shows another infographic for the same (EPA's) data.

4) What percentage of the MSW recycled in 2018 consisted of wood?
5) What percentage consisted of rubber, leather, and textiles?
6) How can the information in Figure 1.11 be misconstrued? In what ways is the infographic misleading?

Compare this infographic with the ones in Figures 1.9 and 1.10 and answer the questions that follow:

7) Which of the three infographics most accurately reflects the data? Why?
8) What are the advantages and disadvantages of each?
9) How can each be misinterpreted?

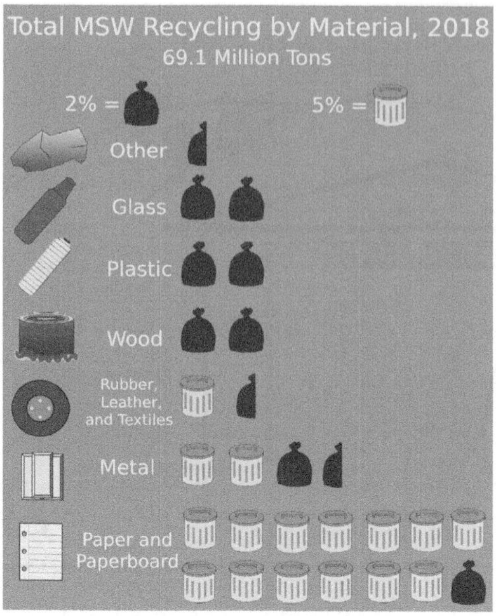

Figure 1.11 An infographic of the EPA's MSW recycling by material data in 2018, based on a pie chart published by https://www.epa.gov/facts-and-figures-about-materials-waste-and-recycling/national-overview-facts-and-figures-materials. Images represent rounding to the nearest whole number. Infographic by Preston Pietruszewski.

Look through some news articles to find examples of misleading infographics and select one to analyze.

10) Describe what you found misleading about the infographic.
11) Do you think the infographic was intentionally misleading? If so, what reasons do you think the designer had for misleading the reader?

IV. What else?

Think about an issue that is important to you. Think of ways to communicate the importance of the issue through an infographic.

1) Why is this issue important to you?
2) Find data related to this topic.
3) Analyze the data using appropriate mathematical tools and methods. You might use descriptive statistics (such as the mean, mode, median, or range) to characterize the data. You might create traditional graphs (such as scatterplots, histograms, or pie charts). Think of how you can make sense of the data and communicate the importance of it.
4) Create an infographic that represents the data you found in a novel way. You can either do this using technology or on paper. Use color and add other details to draw out the important features of your infographic. Your infographic should do the following:

a) Effectively communicate all relevant information.
b) Be mathematically accurate.
c) Be visually appealing.
d) Be creative. Don't just plot points on a scatter plot or histogram. Think about the data and try to come up with a creative way to represent it.
e) Be complete. Presenting selected pieces rather than the whole topic might inadvertently miscommunicate the message to your audience. Ensure that your infographic encompasses the scope of the whole topic.
f) Be drawn to scale. Also, make sure to include legends and labels so anyone reading your graph(s) understands the information being conveyed.
g) Include citation(s) of the data source(s).

Answer each question for your infographic:

5) What information did you hope to convey in your infographic?
6) Why did you choose to design the infographic this way?
7) What challenges did you face while making your infographic? How did you overcome them?

Your design can help you communicate your findings and raise awareness of your chosen issue to your friends and peers. Consider sharing your infographic with a peer and seeking their reactions to it.

V. Story

Regardless of where we see infographics – on TV, in social media, or on posters – they represent other people's interpretations and analyses of data. As consumers of data, we can use mathematical reasoning to critically evaluate and verify their accuracy and overall value.

Just as writing can be used to promote a certain viewpoint, mathematical representations can as well. In fact, people use statistics to deceive others in many contexts. An infographic can further perpetuate these misleading statistics. Undoubtedly, you have encountered advertisements that attempt to sway you to purchase certain goods or services. Often, companies use infographics and other visual representations to bolster their pitch. In the case of advertisements, the consequence of buying one pair of sneakers over another may not be that great, but our decisions about what medicine to buy or what policies to support, for example, can have life-altering impacts.

VI. Going beyond

1) Find an example from the past 50 years of when the public was intentionally misled through an infographic. What was misleading about the infographic?
2) Create a video about the potential misuse of infographics. Include tips for how to be a discerning interpreter of infographics.
3) Pay attention to infographics you encounter in advertisements. Choose one to analyze. Do you think the infographic is accurate or could it be misleading? What do you think was the intent of the advertiser? Do you think the infographic was designed in good faith to accurately convey the data to the public?

4) Research the use of infographics in another academic field. Choose one to analyze. What did you find the most interesting about it? What surprised you? What made you wonder or want to know more?
5) You have been hired as an advertising consultant. Make a list of ethical guidelines for making infographics that your company should follow.
6) Find an infographic about a subject you are unfamiliar with but wish to know more about. What did you learn from the infographic?
7) Go to https://informationisbeautiful.net/beautifulnews/. Which infographics represent mathematics and creativity? Why? Which might be misleading?

VII. Let's reflect

1) Think about the role of infographics in your life. What are the advantages of using infographics? What are their limitations?
2) Keep an infographics journal. Note all the different infographics you encounter over the course of a day (or a week). What do they have in common? In what ways are they different?
3) What did you learn about the issues you identified as important to you through your explorations of infographics in this quest?
4) What steps did you take to ensure that the infographic you made in the 'What else?' section conveyed the data as accurately as possible?
5) Find and describe infographics that prompt you to do the following:

- Consider information in a new way.
- Inspire change in your actions or habits.
- React in an emotional way.

Project 1: Math journaling

In working through the quests in this chapter, you examined the ways in which creativity and mathematics are connected. But what does that mean to you? Are *you* connected to creativity and mathematics? We certainly think so! What worlds (and possibilities) do creativity and mathematics open up for you?

An important part of learning is to internalize what you've noticed, wondered, and explored and to map it against your own interests and previous experiences. One way to do this is through journaling. In this day and age of screens, beeps, and information overload, it is important to take a few minutes a day to sit down and do something just for yourself. One aim of this project is to have you take some time to reflect as well as slow down through journaling. Another aim of this project is for you to reflect with intention about the role of mathematics and creativity in your life, which was the main theme of Chapter 1.

Overview of the project

The main objective of this journal project is to encourage you to make connections between your everyday activities and creativity and mathematics in an intentional way. You can also take this opportunity to reflect on your learning and any new perspectives you might have developed. Most journals entail writing, but this project allows you to tap into your other talents as well. You might choose to create a painting or write a song to submit as entries. You can be creative in selecting different media to present your reflections.

What is the role of creative thinking through mathematics in my everyday life?

This can be the central question that inspires your journal entries. You can choose to answer this question generally in some entries and respond to specific questions, such as the ones posed next, in others. Your individual journal entries might capture what is important to you at the moment, or they might serve to record your reflections over time.

Here are some sample prompts that you might choose to respond to:

1) What is the role of creativity in your life? Do you think differently about creativity now that you've explored the quests in Chapter 1? If yes, write about the differences.
2) Are you convinced that creativity and mathematics are intertwined? Why or why not? If not, what would it take to convince you?
3) Find some piece of art that reminds you of mathematics. The art might be represented in fashion design, music, poetry, or film, for example. Describe what you see/hear/sense. What appeals to you about it? Is there something inherently mathematical about it, or is the connection more personal? Include a photo of the art (if appropriate) or a link to it.
4) Read a recent newspaper article about mathematics. The article should be from within the past six months. Write your reactions to it. Be sure to summarize the article and include the citation.

Again, these are just suggested prompts; feel free to respond to your own prompts or write freely without considering any prompts.

What you will produce

Think outside the bounds of what you envision a typical 'journal' to be. Using technology for your journal allows you to include digital artifacts, such as videos, photos, or music. We hope that you will also consider using an old-fashioned journal notebook. Writing in a notebook gives you a chance to step away from technology. It is more portable than a standard computer (you can write an entry on a hike!), and gives you a different tactile experience. If you are submitting a written document, you might include sketches, cut-outs, photo collages, or poems. Be mindful of what you'd like to get out of the experience; whatever medium you choose, we hope that you can be intentional in making your choice.

Before you get started, determine the duration of your journal project and the frequency of the individual entries. We suggest that you have at least 10 entries in total to ensure that you can reflect on your growth and changes to your perspective over time. Your project might involve one journal entry a week for three months or two weeks of daily journaling, for instance. Whatever the structure of your intended entry, try to jot down your thoughts throughout the day. Note in what situations you are thinking mathematically, creatively, or both. The ideas you 'jot down' – whether in an old-fashioned notebook or using technology – will be the basis of your journal entries.

For each entry, include the date, location (if appropriate), and a descriptive, creative title. You will be submitting one document that includes all of your journal entries.

Chapter introduction: Cited and related works

Beghetto, R. A., Karwowski, M., & Reiter-Palmon, R. (2021). Intellectual risk taking: A moderating link between creative confidence and creative behavior? *The Psychology of Aesthetics, Creativity, and the Arts, 15*(4), 637–644. https://doi.org/10.1037/aca0000323

Ervynck, G. (1991). Mathematical creativity. In D. Tall (Ed.), *Advanced mathematical thinking* (pp. 42–53). Kluwer Academic Publishers. https://doi.org/10.1007/0-306-47203-1_3

Munakata, M., & Vaidya, A. (2013). Fostering creativity through personalized education. *Primus, 23*(9), 764–775. https://doi.org/10.1080/10511970.2012.740770

Torrance, E. P. (1965). Scientific views of creativity and factors affecting its growth. *Daedalus, 94*(3), 663–681.

Quest 1.1: Cited and related works

Csikszentmihalyi, M. (1997). *Flow and the psychology of discovery and invention* (Vol. 39, pp. 1–16). Harper Perennial.

Feist, G. J. (2019). The function of personality in creativity: Updates on the creative personality. In J. C. Kaufman & R. J. Sternberg (Eds.), *The Cambridge handbook of creativity* (pp. 353–373). Cambridge University Press. https://doi.org/10.1017/9781316979839.036

Gotlieb, R. J., Hyde, E., Immordino-Yang, M. H., & Kaufman, S. B. (2019). Imagination is the seed of creativity. In J. C. Kaufman & R. J. Sternberg (Eds.), *The Cambridge handbook of creativity* (pp. 709–731). Cambridge University Press. https://doi.org/10.1017/9781316979839.036

Guilford, J. P. (1950). Creativity. *American Psychologist, 5*(9), 444–454.

Kaufman, J. C., & Beghetto, R. A. (2009). Beyond big and little: The Four C model of creativity. *Review of General Psychology, 13*(1), 1–12. https://doi.org/10.1037/a0013688

Leikin, R., & Lev, M. (2013). Mathematical creativity in generally gifted and mathematically excelling adolescents: What makes the difference? *ZDM Mathematics Education, 45*(2), 183–197. https://doi.org/10.1007/s11858-012-0460-8

Lev-Zamir, H., & Leikin, R. (2011). Creative mathematics teaching in the eye of the beholder: Focusing on teachers' conceptions. *Research in Mathematics Education, 13*(1), 17–32. https://doi.org/10.1080/14794802.2011.550715

Mann, E. L. (2006). Creativity: The essence of mathematics. *Journal for the Education of the Gifted, 30*(2), 236–260. https://doi.org/10.4219/jeg-2006-264

Oleynick, V. C., DeYoung, C. G., Hyde, E., Kaufman, S. B., Beaty, R. E., & Silvia, P. J. (2017). Openness/intellect: The core of the creative personality. In G. J. Feist, R. Reiter-Palmon & J. C. Kaufman (Eds.),

The Cambridge handbook of creativity and personality research (pp. 9–27). Cambridge University Press. https://doi.org/10.1017/9781316228036.002

Runco, M. A., & Jaeger, G. J. (2012). The standard definition of creativity. *Creativity Research Journal, 24*(1), 92–96. https://doi.org/10.1080/10400419.2012.650092

Schoenfeld, A. H. (1985). *Mathematical problem solving*. Academic Press.

Sternberg, R. J. (2019). Everyday creativity. In J. C. Kaufman & R. J. Sternberg (Eds.), *The Cambridge handbook of creativity* (pp. 189–215). Cambridge University Press. https://doi.org/10.1017/CBO9780511763205.013

Sternberg, R. J., & Lubart, T. I. (1995). *Defying the crowd: Cultivating creativity in a culture of conformity*. Free Press.

Torrance, E. P. (1966). *Torrance tests of creative thinking: Norms-technical manual*. Personnel Press.

Zenasni, F., Besançon, M., & Lubart, T. (2008). Creativity and tolerance of ambiguity: An empirical study. *The Journal of Creative Behavior, 42*(1), 61–73. https://doi.org/10.1002/j.2162-6057.2008.tb01080.x

Quest 1.3: Cited and related works

McCandless, D. (n.d.). Beautiful news. *Information Is Beautiful*. Retrieved June 24, 2025, from https://informationisbeautiful.net/beautifulnews

Ritchie, H. (2022, November 30). How many species are there? *Our World in Data*. Retrieved June 24, 2025, from https://ourworldindata.org/how-many-species-are-there

United States Environmental Protection Agency. (2024). *National overview: Facts and figures on materials, wastes and recycling*. Retrieved June 24, 2025, from https://www.epa.gov/facts-and-figures-about-materials-waste-and-recycling/national-overview-facts-and-figures-materials

2 What do you notice?

Now that you have explored mathematics that perhaps challenged your notions about what it means to *do* mathematics, you will use a mathematical lens to view your surroundings. This chapter encourages you to examine mundane objects and everyday situations and extract mathematical meaning out of them. What do you notice? What catches your attention? What do you think is going on? You might add a new dimension to how you view and interact with your environment. In fact, you may never view a strip of paper in the same way again.

One of the tenets of creativity is divergent thinking. Divergent thinking is what some might call 'out-of-the-box' thinking (Guilford, 1950). This kind of thinking is often sparked when we are faced with both a challenge and a **constraint**, such as limited resources. For example, let's say that you are walking in the woods and are faced with an unexpected rainstorm. You don't have an umbrella or rain gear, but you do have the Saran Wrap that was holding your (now eaten) sandwich. You decide to use the Saran Wrap to cover your head. This would be an example where a challenge (walking through a storm) and a constraint (having only Saran Wrap) forced you to be creative. In this chapter, you will examine how constraints can lead you to think creatively about mathematics and how mathematics itself can serve as a constraint as you try to make new connections. We encourage you to be imaginative, flexible, and open to new ideas as you continue to make connections between creative thinking and mathematics.

In our hectic lives and fast-paced environment, it's good to slow down, step back, and take stock of what we notice. In the next three quests, you will make loops from strips of paper, weave together a story out of unrelated items, and go for a stroll and take photos, all with the goal of seeking out mathematics in your surroundings. You will be provided with certain constraints to spur you to make connections, with mathematics as the glue that binds everything together.

Chapter 2: What do you notice?

DOI: 10.4324/9781003541097-3

Quest 2.1: Imagining shapes made from a strip of paper

Creating squares out of loops? No problem!

We use the word 'imagining' in the quest title intentionally. You might associate imagination with whimsy and a sense of dreaminess. Perhaps you think of writing poetry or creating a video when you think of imagination. Imagination can play an important role in mathematics as well. Mathematicians (and mathematics students) use their imagination when picturing shapes in space, taking on different perspectives to solve problems, or looking for patterns in unexpected places. In this quest, you will be imagining the various possibilities that can result from manipulating a simple paper strip.

I. Invitation

This quest invites you to notice your surroundings and engage with creativity through mathematics and develop new mathematical understandings through creativity. We will explore the mathematics made possible by an ordinary object – a strip of paper. You may be surprised by how much mathematics you can explore using a simple strip of paper.

II. What do you notice?

Mathematics can be found in even the most ordinary objects. Take a piece of paper. Actually, let's take more than one piece of paper. Find some standard copy paper (for example, A4 paper). If you have it, color paper works best. Cut your piece of paper lengthwise into about 8 strips. Each strip will be about 30 cm by 2.5 cm. We will eventually need several strips. Unless otherwise instructed, you should attempt to make these strips the same size.

Let's focus on one of these strips.

1) What do you notice about this strip of paper? How would you describe it?
2) What mathematical language can you use to describe it? Generate a list of as many mathematical terms and descriptions that apply as you can.
3) What questions could you ask about this strip of paper?

Now, tape the short edges together, with minimal overlap (see Figure 2.1). Secure the seam by taping both sides of the strip. We will call this a **loop**.

Figure 2.1 A loop made by taping together the short ends of a paper strip.

4) How would you describe this new object?
5) Is there a way to describe it using mathematical language?
6) What are its mathematical features?
7) What new questions arise?

Cut out a new strip with the same dimensions as the previous one, using paper of a different color if possible. Create another loop. Now, you should have two loops of different colors (see Figure 2.2). We will refer to the loops in the photo and in the rest of the quest as gray or white.

Figure 2.2 Two loops: gray (left) and white (right).

Tape the two loops together as in Figure 2.3 so that one loop is oriented 90° to the other. Make sure the tape covers all four seams. We will call the new object that results from taping two loops (in our case, a gray and a white loop) together in this way a **link**.

Figure 2.3 Two loops taped at a 90° angle.

Again, use mathematical language to describe your new object.

8) What new mathematical questions arise?
9) What angles do you see in the link?
10) What features can be measured?

III. Let's see what happens

Take a look at the link you created. Imagine cutting through the **midline** (akin to the center divider of a highway) of the white loop. Figure 2.4 shows the beginning of the cut we are describing. Use it to help you imagine what would happen if we kept cutting. Don't cut yet! For now, just imagine.

Figure 2.4 Starting the cut along the midline of the first (white) loop.

Draw what you think will result after cutting the link in this way.

Now start the cut at any point along the white loop and continue cutting the white loop lengthwise along the midline. Draw or take a picture of the result.

1) What did you notice?
2) What surprised you?
3) How would you describe the result in everyday terms?
4) How would you describe it using mathematical language?

Use your resulting object to imagine cutting the gray strip (which used to be a loop) lengthwise through the midline. Again, don't cut it yet.

5) What do you think will happen? Draw what you imagine.

Cut the gray strip about a fifth of the way (see Figure 2.5).

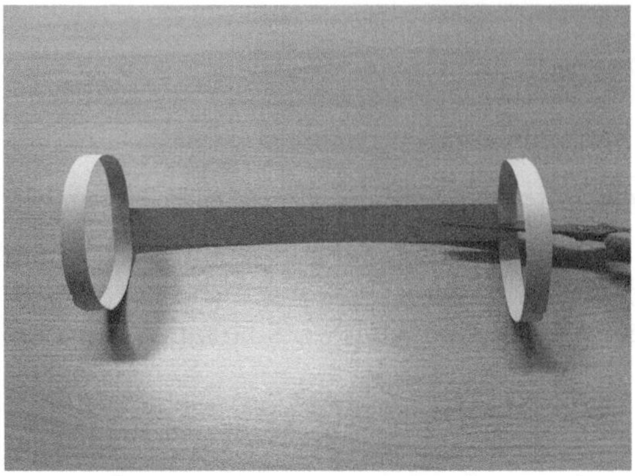

Figure 2.5 Starting the cut along the midline of the second (gray) loop.

6) What do you think will result if you continued cutting? Draw what you imagine.

Now continue cutting all the way along the midline of the gray strip.

7) What happened? Are you surprised with the resulting shape? Not surprised? Jot down your reactions.

Look back at your predictions.

8) Which results did you predict?
9) Were any of your predictions incorrect? Explain why they were incorrect.
10) How would you describe the object you created?
11) What mathematical terms could you use to describe it?
12) What questions could you ask related to the mathematical attributes of your creation? Pose as many questions as you can and jot down possible answers.

IV. What else?

In the previous task, you created a link with two equal-sized loops taped together at a 90° angle, then cut along the midline of each loop. What variations can you introduce to the way we created the link? What can we alter? Generate as many ideas as you can.

Next, we suggest 10 variations. You can pick and choose the variations to try. As you read through them, take note of how you're approaching this task. Do you read one, try it, and then move on to the next one? Do you look for the variations with the longest or shortest description and try those? Do you read them all and then pick a variation to try? It's good practice to be mindful of how you approach challenging problems.

For the variations you chose, what do you think will happen? Doodle what you think might happen and then write a sentence or two explaining each of your doodles. Once you are confident about your predictions, follow the directions for each variation you chose.

Recall that our original paper strip was 30 cm by 2.5 cm.

Variation 1: The gray loop is created with a strip of paper that is half the length (15 cm by 2.5 cm) of the white loop. You attach it at a 90° angle to the white loop. You cut along each of the midlines.

Variation 2: The gray loop is created with a strip of paper that is half the width (30 cm by 1.25 cm) of the white loop. You attach it at a 90° angle to the white loop. You cut along each of the midlines.

Variation 3: The gray loop is created with a strip of paper that is half the length and half the width (15 cm by 1.25 cm) of the white loop. You attach it at a 90° angle to the white loop. You cut along each of the midlines.

Variation 4: Two loops of original size are taped together, not at 90° angles, but at 60° (see Figure 2.6). Be sure to tape all four seams. You cut along each of the midlines.

Figure 2.6 Two loops taped at a 60° angle (for Variation 4).

Variation 5: You tape two loops of original size together at an angle of 90°. Instead of cutting through the midline of the gray loop, you cut ⅓ of the way from the edge of the gray loop (parallel to where the midline cut would be). You then cut the white loop along the midline, as before.

Variation 6: You introduce a half-twist to the gray loop. To do this, you make a loop without taping and flip just the right side edge of the strip. You tape the seams to make a twisted loop (aka a Möbius strip, as you will see in Quest 5.1). You then tape that loop to the white loop (which was constructed without a twist) at a 90° angle (see Figure 2.7). You cut along the midline of each loop.

Figure 2.7 A gray loop with a half-twist taped at a 90° angle to a white loop (for Variation 6).

Variation 7: You repeat the steps in Variation 6, except that this time, you attach the two loops so that they completely overlap where they connect (see Figure 2.8). You can say that the two loops are attached parallel to each other, as opposed to being perpendicularly oriented like in Variation 6. You then cut along the midline of each loop.

Figure 2.8 A gray loop with a half-twist taped parallel to a white loop (for Variation 7).

Variation 8: You introduce a half-twist to each of two loops and attach them at a 90° angle and then cut along the midline of each. Try it twice: Once with the two twists made in the same direction, and then with the two twists made in different directions (one up and one down).

Variation 9: You make a white loop. Then you take a gray strip and cut it halfway along its width, creating two short 15 cm by 2.5 cm strips. You make a loop from each of these strips as usual. You will have three loops total at this point – one white loop and two half-sized gray loops. Then, you tape the two gray loops on opposite sides of the white loop in line with and parallel to it. (Imagine the diameter of the white loop and attach each of the two smaller loops at each endpoint of the diameter, as shown in Figure 2.9.) It is best if you avoid attaching the loops where you have previously taped. You cut through the midline of the white and gray loops.

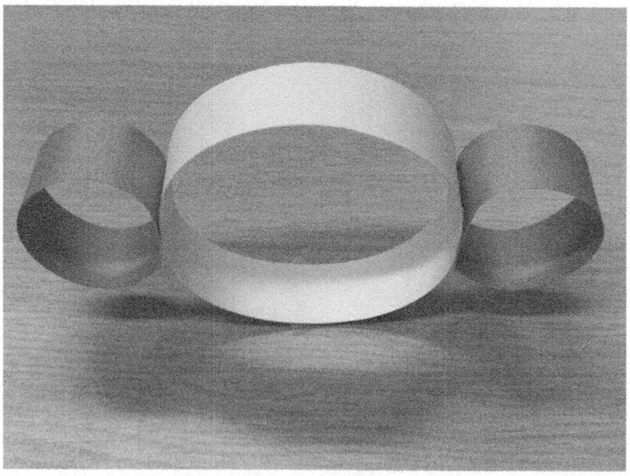

Figure 2.9 Two half-sized gray loops taped parallel to a white loop (for Variation 9).

Variation 10: You make a white loop. Then you take a gray strip and cut it in half, creating two 15 cm by 2.5 cm strips, as in Variation 9. You make loops out of these strips. This time, you tape the two gray loops at an angle of 90°, perpendicular to the white loop (see Figure 2.10). You cut first through the midline of the white loop all the way through where it meets the gray loops and then through the midline of what used to be each gray loop.

Figure 2.10 Two half-sized gray loops taped perpendicular to a white loop (for Variation 10).

V. Story

Think of a mathematics classroom. Think of what it means to learn mathematics. What comes to mind? Often, when people think of mathematics or of learning mathematics, they think of equations, chalkboards, computers, or formulas. Did your image of a mathematics classroom fall along these lines? Has your image of a mathematics classroom changed since the beginning of this book?

Doing mathematics calls for far more than memorizing formulas, computing, and solving equations. It involves imagining, visualizing, conjecturing, noticing, wondering, and creating. It can also involve contorting, cutting, and taping. In fact, making prototypes and using concrete objects to explore patterns is an important part of the discipline.

What you just did in this quest was mathematical. You constructed, imagined, conjectured, and used mathematical language to reason and justify. This is all 'doing' mathematics. Moreover, you discovered something that was most likely new and surprising to you. In fact, we've asked many mathematics professors to predict the result of cutting the loops, and very few could correctly. You engaged in what is considered 'little-c' or little creativity (Kaufman & Beghetto, 2009). The result might not have been earth-shattering, but it was new to *you*. It was a small and personal discovery that engaged you in the creative process of predicting, innovating, and questioning. Your final creations were both mathematical and artistic, and the process by which you created them was creative.

VI. Going beyond

1) Think of your past geometry classes. You'll likely remember explorations of features of shapes such as perimeter, area, and surface area. Use your loops and links as inspirations

for geometric thinking. What questions can you ask about your loops and links? Answer some of your questions.

2) Let's say you wanted to make a picture frame out of links. You have a picture that is 5 cm by 8 cm. Sketch out the frame you would create for your picture. Now, determine the dimensions of the strips of paper you would use to create your frame using the methods you practiced in this quest. Justify your decisions.
3) What new considerations would there be if you were to make a picture frame in this way from paper that was printed with a pattern on one side? How would that change how you make your links?
4) Find a painting or photograph by your favorite artist. What are the original dimensions? Does it have a border (what's called a mat, see Figure 2.11) between the painting and frame? Write directions for creating a border for the artwork using links, specifying the dimensions of the loops.

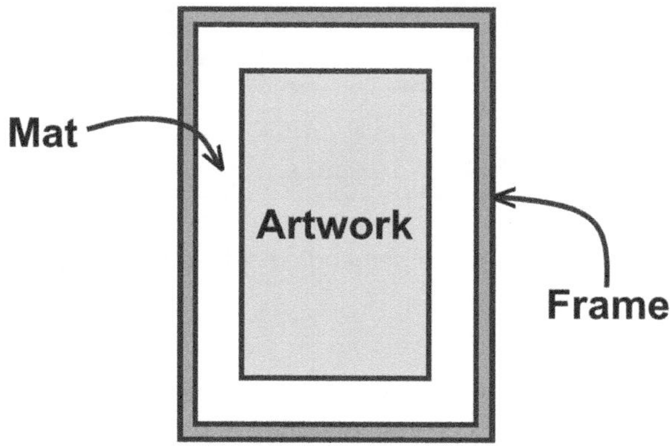

Figure 2.11 A mat surrounding a piece of artwork.

5) How would you make a frame that is a parallelogram but not a rectangle or a rhombus?
6) Make an art piece using multiple cutouts of links and loops. Play with different paper colors and sizes and see where your imagination takes you!
7) For each image in Figure 2.12, make a link from loops and try to create what you see in each image. You may need to attach multiple loops.

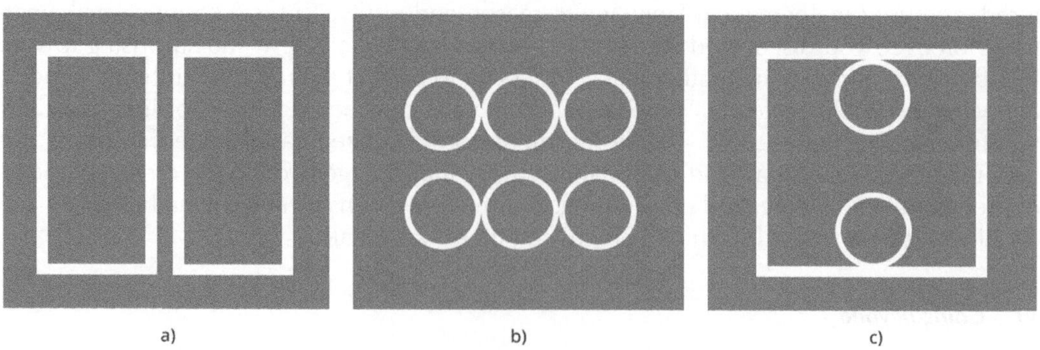

Figure 2.12 Three images for solving Question 7.

VII. Let's reflect

1) Which results in this quest surprised you the most? Which results did you find to be most gratifying?
2) How confident were you in your conjectures and predictions? Were you able to refine your conjectures at some stage of cutting? If so, give an example of when this happened.
3) What was it like to visualize each outcome before cutting? How did visualizing each outcome before cutting make the task easier?
4) Have you ever been asked to conjecture before engaging in a mathematical activity in a different context? If yes, what stood out to you about that experience?
5) Did this quest change your thinking about mathematics? About creativity? If so, how?

Quest 2.2: Connecting ordinary objects to mathematics

What do a slipper, paperclip, and mug have to do with mathematics?

In the previous quest, you were given a strip of paper and asked to pay attention to its attributes. You also studied what resulted when strips of paper were contorted and cut in various ways. In this quest, rather than focusing on a single object, you are going to study collections of everyday objects to seek connections among them and to think about how they might be related to mathematics.

I. Invitation

Consider the purposes of a slipper. Sure, you could wear it, but what else could it be used for? How many uses can you list? Of the uses you can think of, which are the most interesting? Which are the most outlandish? Which are the most useful or practical? What makes these uses creative? What about a paperclip? What are some creative uses for a paperclip?

Similar questions that ask you to generate as many uses for an object as possible appear on a 'test' of creativity that measures your divergent thinking skills (Guilford, 1950). Divergent thinking is one of the hallmark features of creative thinking and involves multiple components. It refers to thinking of as many solutions to a problem as possible. It also includes making connections among ideas that may at first seem unrelated. In fact, under some definitions of creativity, higher levels of creativity are called upon when the association is less predictable (Tromp, 2022). Someone who engages in divergent thinking will also change course as necessary. This way of thinking differs from many people's notions of what it means to engage with mathematics – that there is only one correct answer, as well as one correct approach to a mathematics problem. Both divergent and convergent thinking, which involves narrowing down the possibilities, are important to creativity. In this quest, we will focus on divergent thinking.

We challenge you to engage in divergent thinking to make connections not only between objects and their possible uses but also between objects and mathematics.

1) How can mathematics be used to make connections among seemingly unrelated everyday objects?
2) How can mathematics be used to tell a story?
3) How can objects be used to convey mathematical thinking?
4) In what ways is the world around us mathematical?

II. What do you notice?

When you see a bird, you might notice its colors, size, or sound. Or, you might observe the bird through a more mathematical lens. You could become curious about its flight path and what a graph of the bird's height at various points along its path would look like. You might also wonder about the cluster of trees that the bird visits. Does the bird fly to the closest tree? Does it seem to choose a particular species of tree? Is there a pattern? How can mathematics help us make sense of the bird's movements and apparent decisions? In this quest, you will notice things in a different way as you view objects through a mathematical lens.

Instead of focusing on one object (or on a bird, as we did in the example), you will work with five objects that at first seem unrelated. If you are not in a classroom setting, find five objects from your surroundings. One way to choose unrelated objects is to choose them from different rooms or places (one from the kitchen, one from outside, another from your bag, etc.). We are refraining from giving you examples because we do not want to influence your choices too much!

Jot down all of the connections you see among your five objects. You might connect two objects because they share a similar feature or use, or you might find a characteristic that connects all of your objects. Keep track of how you are organizing your ideas. You might jot down your thoughts, create a diagram, or use technology to organize your thoughts in a creative way.

1) What features are you considering for your chosen objects? Are you considering the objects' uses? Shapes? Textures? Or perhaps you are considering more abstract features such as what they symbolize or some literary connection.
2) Which object was the most difficult to connect to the other objects? Why?

III. Let's see what happens

Now that you've made connections among your five objects, we challenge you to make connections between your collection of five objects and mathematics.

Study each object in your collection

1) What about each of your objects makes it mathematical? The connections you make between each object and mathematics might not be initially obvious, but you can engage with divergent thinking as you consider the different possibilities for how each object is somehow connected to mathematics.

Consider the five objects together

2) What mathematical features or characteristics do the objects share?
3) Pose some questions related to the objects' shared mathematical features or characteristics. Be sure to consider all five objects together.
4) How might you answer those mathematical questions?

Create your model

Design a model or structure with the five objects you were given (or found). See Figure 2.13 for an example of a structure designed with the following objects: playing cards, a cup, table tennis balls, eyeglasses, and string. Can you see what connections to mathematics this student tried to represent?

Make your own model with your five objects. Your model should be both informative and purposeful. Think carefully about each part of your design before assembling the model to ensure that it clearly communicates your mathematical thinking. Feel free to use tape, string, or writing implements to combine objects, make marks on your objects, or make certain features stand out. You can also draw a schematic that makes your connections clear. Be prepared to explain your model to others.

Consider the process you used to make these connections

5) List all of the mathematical connections you initially came up with, even if they ended up not being included in your model. How did you narrow down your ideas?

Figure 2.13 A sample mathematics model connecting five ordinary objects. Model and photograph by Adam Leszczyński.

6) Which object was the most difficult to connect to the mathematics of the other objects? Why?
7) In what ways did you engage in divergent thinking during this activity?

IV. What else?

So far in this quest, you have connected five objects to each other and to mathematics. These objects were either provided to you or chosen by you. Now, consider the five objects listed in Table 2.1.

Table 2.1 A table for recording possible mathematical connections to objects.

Object	*Mathematics topics*	*Possible connections*
Slipper		
Paperclip		
Mug		
Cactus		
Skateboard		

The table contains three columns, labeled 'Object', 'Mathematics topics' and 'Possible connections'. Copy this table, provide examples of mathematics topics that you think are connected to each object, and describe each connection in the last column. Remember – there is no 'right' answer! List as many connections as possible.

What was the process of making these connections like for you? Did you feel the need to be creative? One way to engage in creative thinking is to challenge ourselves to make connections when they are not immediately apparent. We often make these types of connections in our day-to-day lives, even if we are not aware that we are doing so. For example, if you are hanging a picture on a wall and can't find a hammer, you might think to use a pan to 'hammer' in the nail. In this case, you would have made a connection between the pan and its potential use as a hammer.

In this exercise, instead of finding an alternative use for an object, you were asked about how it connects to mathematics. The object and the instruction to connect it to mathematics served as constraints that focused your thinking.

V. Story

Creativity is often heightened by the presence of constraints. As we mentioned in the chapter introduction, constraints are limiting factors such as rules, available resources, or capacity. They help us channel our creativity in a focused way and challenge us to think in unconventional ways. Studies have shown that levels of creativity increase when constraints are introduced (e.g., Medeiros et al., 2018). Constraints help us focus our thinking and challenge us to think of atypical connections (Tromp & Baer, 2022). The objects you were given (or found) for this quest, together with the requirement of using all and only those objects, represent those constraints. It may have seemed at first that your objects were not connected to mathematics, but we challenged you to make those connections and encouraged you to think about the objects and mathematics deeply – and to use creativity to connect the two.

Constraints play an important role in formal mathematics as well. For example, when writing proofs, you are constrained by certain assumptions. Likewise, when trying to create a mathematical model of a situation, you are bound by constraints. When a mathematician is helping a company develop an algorithm for maximizing profits, for example, there may be constraints such as environmental considerations, employee capacity, or a deadline. These constraints force the mathematician to think creatively to come up with a feasible plan and fulfill the company's needs.

One famous assertion whose unassuming constraints led to a centuries-long debate was Fermat's Last Theorem. Written in the margin of a book in the 1630s, Pierre de Fermat made a conjecture that $a^n + b^n = c^n$ had no positive integer solutions when n is greater than 2. This problem is constrained by the fact that a, b, and c have to be positive integers and the three exponents have to be the same integer greater than 2. Although this seems like a very simple theorem, it took over three centuries for it to be proven until the English mathematician Andrew Wiles published his proof in the mid-1990s.

VI. Going beyond

1) In this quest, you made connections between a set of objects and mathematics. Identify five topics from another subject area (for example, biology, psychology, theater, arts) you are studying. Consider these to be 'objects'. What connections can you make between each of these objects (topics) and mathematics?
2) There are examples in history of unexpected connections made by scientists that led to significant discoveries. For example, you may have heard about how a falling apple inspired Isaac Newton to explore principles of gravity. Find out about other discoveries that were made through creative connections. Make a video or write a short story about your findings.

3) Keep track of constraints in your day-to-day activities. Do they encourage you to be creative? In what ways?
4) There are many constraints involved in playing sports. Think of your favorite sport and make a list of all the ways in which constraints affect the sport. For example, what would it be like to play basketball without a well-defined court?
5) In this quest, you made connections between objects and mathematics. What connections can you make between objects in a specific environment and mathematics? For example, go to a local botanical garden. Describe the connections you see between the plants or other features of the garden and mathematics.
6) Connect objects to mathematics by creating a mathematical game with objects of your choice.

VII. Let's reflect

1) To make a model or structure with five seemingly unrelated objects in this quest, you were asked to first notice them as being mathematical objects and then to connect them as a group to mathematics. How did you approach this task? What felt natural about this experience? What did you learn in the process?
2) How does this quest help you think differently about mathematics?
3) How did this quest differ from your past experiences with mathematics?
4) Do you agree that constraints encourage creativity? Do they always? Give examples to back up your arguments.
5) Think of a situation in your life when a constraint forced you to be creative. Describe the situation and how you dealt with the constraint. What did you learn from the experience? Were you able to 'do more with less', as the saying goes?
6) What was the most creative connection you made between your objects and mathematics in this quest? What makes this connection especially creative?

Quest 2.3: Connecting mathematics to ordinary objects

Let mathematics guide you!

In the previous quest, you were given a set of objects and asked to notice the mathematics that ties them together. Having to use that set of objects acted as the constraint that spurred you to be creative and encouraged you to tap into your mathematical knowledge. In this quest, we flip the script. You will be given a list of mathematics topics and be asked to connect the topics to what you see around you.

This quest will encourage you to go outdoors. You will first choose a list of mathematics topics and try to find representations of those topics in your surroundings during a 'math walk'. You will then create an interactive map that showcases the connections that you noticed.

I. Invitation

Mathematics is not restricted to what is presented to us in textbooks. Look around you. You will undoubtedly see many objects with mathematical characteristics. Consider the photo of a walkway in Figure 2.14. In just this one photo, you can find line segments, angles, perspective, and area, just to name a few mathematical ideas.

Figure 2.14 A Meadowlands walkway in Lyndhurst, New Jersey.

II. What do you notice?

Consider the Dol Hareubang ('stone grandfather') statue from Jeju Island, South Korea, in Figure 2.15. There are many similar statues on the island: The oldest ones are believed to be about 500 years old. These statues vary in size and can be up to 3 meters in height. At first, you may notice the statue's shape, color, or surface texture. What are some other features you notice in the statue? What if we asked you to note features of the statue that are related to specific mathematics topics?

Figure 2.15 A stone Dol Hareubang statue from Jeju Island, South Korea.

Image by Noppasin Wongchum, retrieved from Shutterstock https://www.shutterstock.com/image-photo/dol-hareubang-statue-jeju-do-sanbangsan-410581348?trackingId=1718df3f-f052-4f2d-b372-045a14390dac

Table 2.2 should look familiar to you from Quest 2.2. Recall that in that quest, we provided a list of objects and asked you to identify mathematical connections to each. In this table, we provide you with five mathematics topics. Can you identify features of the statue that connect to each of these topics?

Table 2.2 A table for recording features of the Dol Hareubang statue that connect to mathematics topics.

Features of the statue	*Mathematics topic*	*Possible connections*
	Fractions	
	Lines	
	Probability	
	Statistics	
	Angles	

The table contains three columns, labelled 'Features of the statue', 'Mathematics topic', and 'Possible connections'. Copy this table, list features of the sculpture that are connected to each mathematics topic and describe each connection in the last column. Remember – there is no 'right' answer! List as many connections as possible.

Let's linger on the statue a bit longer ... .

Refer to the statue in Figure 2.15 to answer these questions:

1) What features of the statue make it interesting or artistic to you?
2) What percent of the statue is the 'head'? How did you determine this?
3) Suppose the statue is 3 meters tall. If you were to make a miniature model of this statue, what scale factor would you use and why? What information would you use? What assumptions would you make?
4) Where is the center of the statue? How are you defining 'center'?
5) Dol Hareubang statues are made from basalt, a volcanic rock.
   a) How much do you think this statue weighs?
   b) How much would it cost to make this statue today?
   c) How did you determine these amounts? What additional information did you need?
6) What other mathematical ideas or processes do you see evidence of in the statue?
7) What characteristics of creativity does the statue represent?

III. Let's see what happens

The Dol Hareubang statue is likely too far away for most of us to view in person, but we do not need to travel far to see mathematics around us. In this quest, you will seek mathematics in your surroundings closer to home.

You are going to select three mathematics topics, go for a walk, and then create an interactive map of photos that showcases those topics. For each photo, you will also write a task or pose a question. For the purposes of this quest, we define an interactive map as a medium through which you can pose mathematical questions based on locations in your town to the reader.

Let's say that you selected geometry, statistics, and rate of change as your mathematics topics. You then went for a walk around campus. You might create a map like the one in Figure 2.16.

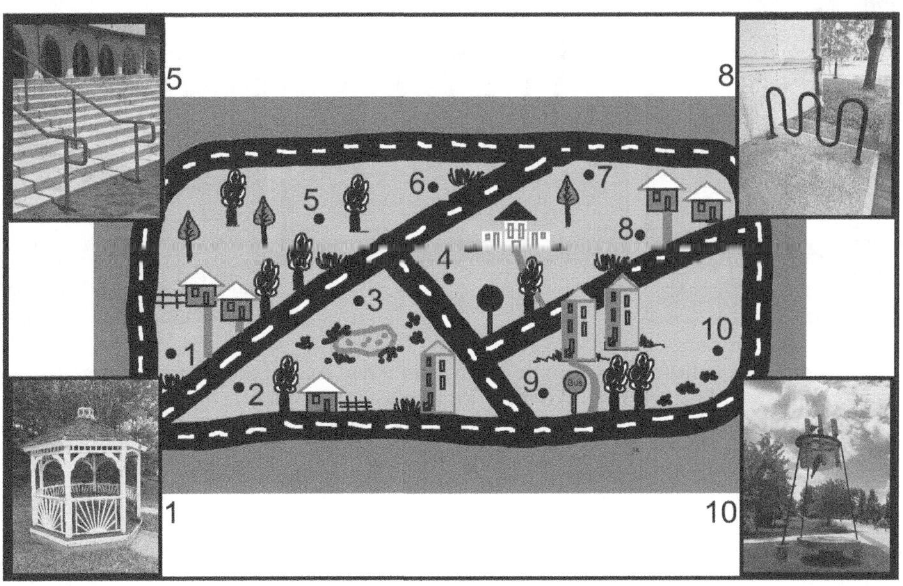

Figure 2.16 An interactive map of ten locations (with four sample photographs, photograph 10 taken by Mac Adams).

On your walk around campus, you noted several locations where you saw connections to at least one of these mathematics topics. You would have then written tasks based on these locations. For example, for Photo 1 in Figure 2.16, you might have developed the following task connected to geometry:

> This gazebo has a wooden base in the shape of a regular octagon; that is, the base is a polygon with eight equal sides. Its side panels form a 'wall' around the base that reaches up to half of the gazebo's height. You are hired to design a bench for the inside of this gazebo. What shape would you need the bench to be in order to maximize the number of people that could comfortably sit inside the gazebo? What kinds of measurements and tools would you need to design the bench? How would you determine the maximum number of people that could fit on the bench?

Let's practice a bit more, this time with rate of change. Figure 2.17 shows the three other photos included on the interactive map.

1) Where in each photo do you see connections to rate of change?
2) Write a task (or pose a question) focusing on rate of change
    a) Based on only one of the photos.
    b) That connects two of the photos.
3) For an added challenge, write a task that incorporates all three photos.

Figure 2.17 Three objects from the example of an interactive map. Left photo: A staircase with metal railing. Middle photo: A bike rack. Right photo: *Ophiuchus the Serpent Bearer* sculpture created in 1988 by Mac Adams, assisted by Dr. Mary Lou West. Photograph on right by Mac Adams.

IV. What else?

Make your own interactive map

Choose at least three mathematics topics from the list that follows, challenging yourself by choosing ones that at first seem unrelated to each other. Some of these topics are more general than others. You can also choose other mathematics topics that are interesting to you. Feel free to use the Internet to brush up on some of these areas of the field of mathematics.

- Exponential growth
- Rate of change
- Statistics
- Quadratic functions
- Trigonometry
- Geometry
- Acceleration
- Properties of polygons
- Proportions
- Hypothesis testing

Now that you've selected three mathematics topics, you are going to go for a 'math walk'. This can be a real walk around your neighborhood or a virtual walk of a place you've always wanted to visit using a digital map application. Your list of three mathematics topics will guide you on your walk. Your main challenge will be to note and photograph (or screenshot if you're taking your walk virtually) at least five locations (or objects) that showcase your three mathematics topics. These notes, photographs, or other recordings (such as audio) may be of what you see, what you hear, or what the place represents. Make sure to find at least one location for each topic. For example, if one of your topics is exponential growth, you might take a photo of a bank you walk past because you thought of the relationship between interest rates and exponential growth. Also on your walk, you might take a picture of a tree branch that splits into two smaller branches at each juncture. This might remind you of doubling, which is also a type of exponential growth.

- Bring a notebook, a pen or pencil, and a phone (or camera) to record each location. You can also bring a measuring tape or other measuring tools.
- Take photos of at least five locations from your walk that connect to your mathematics topics, making sure that you have at least one example for each of your three topics.
- You will need to take detailed notes on each object and jot down any details one might need to complete your task (or answer your question).

For each of your five locations, you will create a math task to challenge your classmates.

- Write a task related to the mathematics you see. For example, perhaps you come across a plaque with two circular motifs. You might pose the question, "Are all circles proportional?" You would jot down your observations about the plaque, including measurements and any special features related to the circles.

Once you have your five photos and tasks, create an interactive map. Insert your photos and mark their respective locations on the map. Specify at least one mathematics task or question for each photo. Your tasks can be printed out or shared via a link. You might choose to use poster paper or create a digital map (even a dynamic digital map). Be sure to provide a rich description of the mathematics represented in each photo so that the reader has all of the information they need to complete your task.

Your map does not have to be drawn to scale as it won't be used to determine distance or direction. Rather, the purpose of your interactive map will be to ask thought-provoking questions based on interesting locations in your area. A key of possible solutions should be provided in a separate document.

This is your opportunity to connect creativity, mathematics, and art, so make your map colorful and visually appealing!

V. Story

In Quest 2.2, you identified the mathematics of seemingly unrelated objects. That is, given a set of objects, you found their shared mathematical connection(s). In this quest, you were asked to identify objects that matched preselected mathematics topics. When given a constraint, whether unrelated objects to connect to mathematics in Quest 2.2 or mathematics topics to connect to objects in this quest, you made connections in situations where you perhaps did not see connections before.

Let's look at the Vessel building in New York City (see Figure 2.18). This visually stunning structure showcases geometry in a beautiful way. What if we consider more than what can be seen through the lens of geometry? We can make connections between the Vessel structure and functions, symmetry, iterative patterns, statistics, probability, and tessellations, just to name a few topics. Do you see these connections?

For example, let's say that two of your chosen topics were probability and statistics. You can see in Figure 2.18 that the area around the structure is quite crowded. You might ask the following questions related to probability: What is the probability of being able to enter the Vessel within five minutes of your arrival if you go in the morning? Evening? New Year's Eve? What policies should the management adopt to ensure that entrants are safe and comfortable during their visit? How does statistics play into these decisions? Use this same photo to try a similar exercise for another topic in mathematics.

Figure 2.18 The Vessel structure in New York City. Photograph by Bryant's Juarez, retrieved from https://www.pexels.com/photo/brown-and-white-architectural-building-7839630/

VI. Going beyond

1) Name three different objects that come to mind when you encounter each of the following words:
    a) Line
    b) Perimeter
    c) Median
    d) Pi (π)
    e) Proportions
2) Choose a mathematics topic. Make a collage of structures, statues, and spaces in your area that you can connect to this topic.
3) Choose one of these mathematics topics and generate a list of 'real world' connections of that topic:
    a) Quadratic functions
    b) Exponential functions
    c) Square roots
    d) Derivatives
    e) Probability
    f) Absolute value
4) Give an example of a mathematics topic for which you think there are no 'real-world' connections. First explain why you think there are no connections to this topic then do some Internet research to see if that is in fact true.
5) Your seventh-grade cousin is working on homework on absolute value. Create a video for your cousin showing them the connections between absolute value and their life.
6) Consider a topic from your major (for example, history, chemistry, linguistics), like you did for the mathematics topics in this quest. Take a walk outside or visit a neighborhood virtually to find objects connected to your topic. Create an interactive map for that topic.

VII. Let's reflect

1) Was there a mathematics topic that was easier to see evidence of in your surroundings than others? Was there one that you could not find a connection to?
2) What other field(s) of study, outside of mathematics, do you frequently see connections to in your surroundings?
3) What from your collection of objects in this quest had the most unexpected mathematical connection? What makes this connection so special?
4) In what ways were you being creative when finding objects connected to particular mathematics topics?
5) Think of a mathematics topic you struggled with in school. What are some ideas or situations outside of mathematics that connect to this topic? (You might do some Internet research on this.) Did learning about these connections help you better understand the topic now?
6) Did the work you completed for any of your chosen mathematical constraints change your view of the objects around you? Explain your response.
7) Compare and contrast your experiences of finding shared mathematical connections across different objects (Quest 2.2) and finding objects that connect to the same mathematics topic (Quest 2.3). Was one easier or harder for you? Why do you think that is?

Project 2: Math photo project

This project challenges you to be on the lookout for mathematics around you and to capture those moments through photography. You will describe and investigate the mathematics that appears in the scene or object that you photograph.

1) Take a photograph of evidence of mathematics around you. The photograph must be

   a) Taken by *you*
   b) Of the actual object (it cannot be of an image you found in the newspaper or online, for example)
   c) Not altered using photo editing software (see exceptions)
   d) In high resolution.

   The photograph may be

   a) In color or black-and-white
   b) Of an animate or inanimate object
   c) Edited with the use of editing software if all you are doing is adding lines, pointing out features, or adding text.

2) Select one photograph to analyze and submit the following:

   a) The title of your photograph. Think of a creative title that will convey the mathematics you noticed as well as the specific phenomenon you captured.
   b) The date and location of the photograph taken.
   c) A description of what the mathematics in the photo means to you. This should include responses to the following prompts:

      i) What are the mathematical ideas in the object of your photo that you noticed?
      ii) Describe or point out what you observed.
      iii) Generate a list of questions you can ask about the mathematics in the photo.
      iv) Select one of your questions from iii) and refine it to make sure that it is phrased clearly.
      v) What mathematical tools or specific knowledge will you need to answer your chosen question? Jot down what you wish you had at your fingertips. Perhaps you think you would need a tape measure or barometer, or perhaps an advanced understanding of physics would help you.
      vi) Describe how you might approach your problem. Try to combine your mathematical knowledge with whatever other information you can find. If possible, find an exact answer to your question or estimate a solution.

Here is one example and analysis. Note that we present just one possible way to use this to investigate the mathematics in the photo. As a challenge, provide another!

- Title: Whirlpool of Mathematics
- Taken in Tokushima, Japan. May 27, 2024
  - I took this photo of whirlpools (see Figure 2.19) at a famous spot called 'Naruto', in the Inland Sea (Setonai-kai) of Japan. The photo was taken from a bridge. The whirlpools here are created by a sudden drop in the sea floor. There were many of these whirlpools occurring in the sea. The photograph shows just a few. Seeing that the whirlpools

Figure 2.19 A photograph of a Naruto whirlpool, Japan.

differed in size made me wonder: Is the average diameter of whirlpools in a particular area related to the depth of the sea floor at that location?

- I would need access to the depth charts for different areas where the whirlpools are evident. I would then photograph several whirlpools occurring at around the same time in different areas of different depths of the ocean floor. I would want to include whirlpools of different sizes. I would need to decide what I consider the diameter of a whirlpool and measure the diameter of each as seen in the photographs.
- To identify the whirlpools, I will focus my attention on the areas of the photographs that are more white than gray. I will consider that to be the central part of each whirlpool. I would then measure each and compare against the depth of that location.
- On the *xy*-plane, I could graph the depth of the sea at that location as the independent variable (*x*) and the average diameter of the whirlpool as the dependent variable (*y*) to determine the relationship between the depth of the sea and the diameter of the whirlpools, if there is one.

Once you have analyzed your photo, we encourage you to share your photo with someone else: a peer, a friend, or a family member. See what evidence of mathematics they notice in your photo. Perhaps, together, you can come up with other mathematical questions to ask. You could even hold an interactive mathematics photo exhibit with your peers or start a conversation on social media about taking photos of mathematics around us. See where that conversation takes you!

Chapter introduction: Cited and related works

Guilford, J. P. (1950). Creativity. *American Psychologist, 5*(9), 444–454.

Quest 2.1: Cited and related works

Kaufman, J. C., & Beghetto, R. A. (2009). Beyond big and little: The Four C model of creativity. *Review of General Psychology, 13*(1), 1–12. https://doi.org/10.1037/a0013688

Vishnubhotla, M., & Munakata, M. (2018). The mathematics of loopy pictures frames. *Teaching Mathematics in the Middle School, 23*(4), 231–234. https://doi.org/10.5951/mathteacmiddscho.23.4.0231

Quest 2.2: Cited and related works

Guilford, J. P. (1950). Creativity. *American Psychologist*, 5, 444–454.

Medeiros, K. E., Steele, L. M., Watts, L. L., & Mumford, M. D. (2018). Timing is everything: Examining the role of constraints throughout the creative process. *Psychology of Aesthetics, Creativity, and the Arts, 12*(4), 471–488. https://doi.org/10.1037/aca0000148

Tromp, C. (2022). Integrated constraints in creativity: Foundations for a unifying model. *Review of General Psychology, 27*(1), 41–61. https://doi.org/10.1177/10892680211060027

Tromp, C., & Baer, J. (2022). Creativity from constraints: Theory and applications to education. *Thinking Skills and Creativity, 46*, 1–11. https://doi.org/10.1016/J.TSC.2022.101184

Quest 2.3: Cited and related works

Munakata, M., & Vaidya, A. (2012). Encouraging creativity in mathematics and science through photography. *Teaching Mathematics and Its Applications: An International Journal of the IMA, 31*(3), 121–132. https://doi.org/10.1093/teamat/hrr022

3 Where is the mathematics in play?

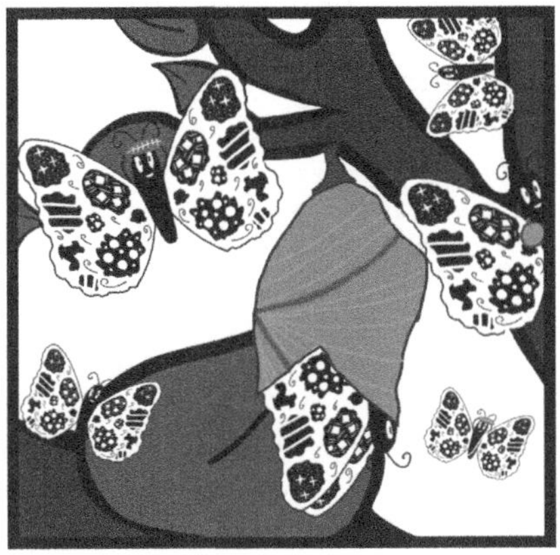

In the previous chapters, you summoned your creativity to interact with your surroundings as viewed through a mathematical lens. In this chapter, you are going to consider how mathematics relates to certain activities associated with play and recreation. The title of this chapter is 'Where is the mathematics in play?', but it could also be 'Where is the play in mathematics?' You will be thinking of mathematics as you play and also be playful in your approaches to mathematics.

When we think about how we use mathematics in our daily lives, we might think of chores such as cooking or balancing our budget, but we also encounter mathematics when we are relaxing and having fun. For example, there is mathematics inherent in many recreational activities including playing games, taking part in sports, and playing with toys. In fact, recreational mathematics – which involves many types of mathematics in its analysis of 'play' – is in itself a bona fide branch of mathematics. With regard to creativity, it is no accident that many of the characteristics of creativity overlap with characteristics of play: Both play and creativity require self-motivation, are guided by mental rules but leave room for improvisation, are focused on the process more than the outcome, and are imaginative. All of these characteristics come out best when one is not in a stressed situation (Burghardt, 2011).

Creativity also comes into play (pun fully intended) when exploring the connections between mathematics and play. We will demonstrate to you that this exploration is inherently creative. When using mathematics to analyze play, you are engaging in non-algorithmic problem solving (Ervynck, 1991), being flexible (Torrance, 1996), and seeking personally meaningful insights (Kaufman & Beghetto, 2009). Play is also involved when engaging in mathematics. As you've seen in previous quests, this book embraces playful, non-procedural approaches to mathematics including doodling, contorting paper, and imagining outcomes.

Be prepared for some messiness. You will be tossing things in the air, dipping your fingers in finger paint, and scattering gummy bears about. Being creative and doing mathematics can be both playful and messy!

DOI: 10.4324/9781003541097-4

Chapter 3: Where is the mathematics in play?

Quest 3.1: Using mathematical notation to make sense of juggling

Do you feel the rhythm of mathematics?

We all 'juggle' things from time to time. For example, we juggle our obligations, classes, and commitments. Juggling is often used as a metaphor for attending to multiple tasks at once. For this quest, however, we will be less metaphorical and more literal. We will 'play' with mathematical ideas by juggling or tossing objects in the air and catching them in a continuous and rhythmic pattern.

I. Invitation

Find at least three objects of the same size and shape to juggle. You might use bean bags, juggling balls, or tennis balls (although good luck chasing after them!) Ideally, the three objects would be of different colors, but it's okay if they're not. Even if you are a confident juggler, we do not recommend juggling raw eggs (for obvious reasons!)

Juggling doesn't come easily. It takes lots of practice. But don't worry – you don't need to be able to juggle to complete this quest. Simply tossing a bean bag or a tennis ball will give you a sense of the rhythm that is necessary to juggle and will no doubt lead to a huge appreciation for people who can juggle four, five, or even ten balls. Regardless of our juggling prowess or lack thereof, we can analyze the patterns that emerge and the mathematical structures of juggling.

The diagram in Figure 3.1 depicts a juggling pattern.

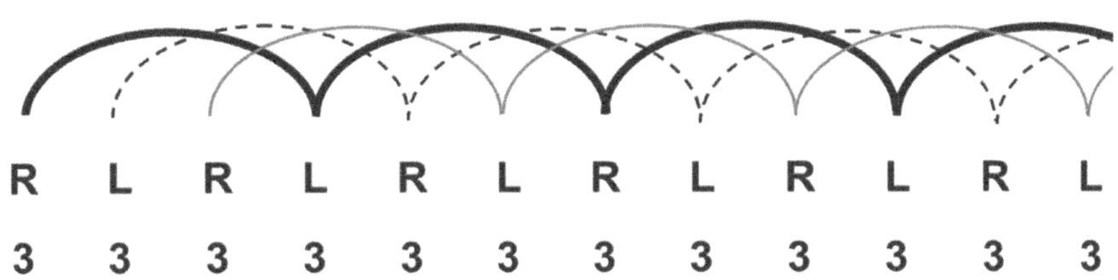

Figure 3.1 A visual representation of a juggling pattern.

1) What about the diagram suggests that it is about juggling?
2) What do you notice about the diagram? What conjectures can you make?
3) What do you think the juggler is doing?
4) How many balls are being juggled?

II. What mathematics do you see?

Start with one ball. Toss it in the air from one hand to the other. Toss it until you get into a rhythm and can toss and catch it consistently. Try to toss it so that it reaches about the same height each time and so that your hands are moving to a constant beat. Counting out loud might help you achieve this constant beat. Considering that you will be adding more balls to juggle soon, be mindful of how you are tossing a single ball and keep track of what is working and what is not.

1) What is involved in juggling one ball with both hands? What can you describe or keep track of? List all of the aspects of juggling one ball that you notice.

2) What mathematical questions come to mind from juggling one ball?
3) Answer each of your mathematical questions with a doodle or a diagram. You might use color, different notations – whatever helps you keep track of the different attributes of juggling one ball.
4) How would you describe the movement of your hands?
5) Are there other ways to juggle one ball? Try to vary how you juggle a single ball in at least three different ways.

III. Let's see what happens

In considering juggling one ball, you might have thought about mathematical attributes such as the height the ball reaches, the timing of your throws, or the rhythm of your throws. These are all elements of juggling we will study as we analyze the mathematics of juggling.

We are going to focus on 3-ball juggling where each ball crosses from one hand to the other across the body, making an 'X'. This is perhaps the pattern of juggling that first comes to most people's minds. This is called a **3-ball cascade** and is different from what is called the **3-ball shower**, where the balls go around in a circle clockwise or counterclockwise. The 3-ball cascade makes an 'X' across your body, whereas the shower makes a circle. Find a video of each of these juggling patterns and convince yourself that they are different. Take note of the differences you see.

Rewatch your video on 3-ball cascade juggling. Try juggling three balls in this way so that each ball crosses your body and they all reach about the same peak height. This is easier said (or written) than done. Even if you can juggle just a few throws at first, that would be quite an accomplishment.

Let's focus on the hands. (You might rewatch the video in slow-motion for this.)

1) What patterns do you see in the movement of the hands?
2) Represent this on paper. What diagram might you draw to represent what is happening with the hands?

Now, focus on just the balls.

3) What patterns do you see in the movement of the balls and where they land?
4) Represent this on paper. Can you keep track of each ball? What diagram might you draw to represent what is happening with each of the balls?

Siteswap notation

SUPPORT MATERIAL *Watch video 1.* This video features Sean Gandini and Kati Ylä-Hokkala. They are world-renowned artists and performers who founded Gandini Juggling in 1992. In the video, they set the stage for analyzing juggling using mathematics. We will be working with **siteswap notation**, which is a particular way of recording juggling patterns. Siteswap notation is a sequence of numbers that represent how many beats there are between when each ball is thrown and when it will next be caught and rethrown.

As you watch the video, note how each throw of a ball counts as one beat. If necessary, slow down the playback speed on the video to convince yourself of this.

Keep in mind the assumptions necessary for the siteswap notation:

- Hands are always alternating: right, left, right, left . . .
- Throws are at equally spaced intervals.
- If a ball arrives in a hand, the ball has to leave, or be thrown from, that hand.

Rewatch Video 1 and see if you can follow one ball at a time. For example, how would you describe what is happening with just the red ball?

Watch video 2. In this segment of the video, Sean shows us the siteswap notation for the 3-ball cascade. Pay attention to the balls as they are being thrown. Because each ball (red, green, white) is being rethrown every three (3) beats, the siteswap notation would be (3, 3, 3). Because the 3 is repeating for each color, the siteswap notation can also be denoted simply as (3).

In the video, we are also introduced to a notation that allows us to follow each of the three balls as it is being juggled. This is called a **space-time diagram** (see Figure 3.2). In our diagram, line styles used for the arcs (for example, dashed or solid) indicate different balls.

A space-time diagram uses the numbers from siteswap notation along the bottom (concerned with beats) together with the labels 'R' and 'L' (to designate the right and left hands) and arcs to track the movement of each ball over time. Each arc starts with the hand from which the ball is thrown and ends with the hand with which it is caught and immediately rethrown. The numbers indicate how many beats later the ball will land (and immediately be rethrown).

Rewatch Video 2. Pay close attention to the part of the video where Sean describes the space-time diagram for the 3-ball cascade. Try to track the path of each ball as you follow the diagram. Keep in mind that time is progressing as we read from left to right. Even though the balls look like they are rolling from the left side of the diagram to the right, this is not the case. We are asking you to do a little bit of mental 'juggling' here!

Let's continue to try to make sense of the diagram in Figure 3.2.

5) Write a couple of sentences about what a space-time diagram shows.
6) Write a couple of sentences about what a space-time diagram does not show.
7) Compare the diagrams you developed earlier (before you learned about the siteswap notation and space-time diagrams) with the space-time diagram shown in Figure 3.2.
   a) How are they similar?
   b) How are they different?
   c) What information does your notation capture that the space-time diagram doesn't?
   d) What information does the space-time diagram capture that yours doesn't?

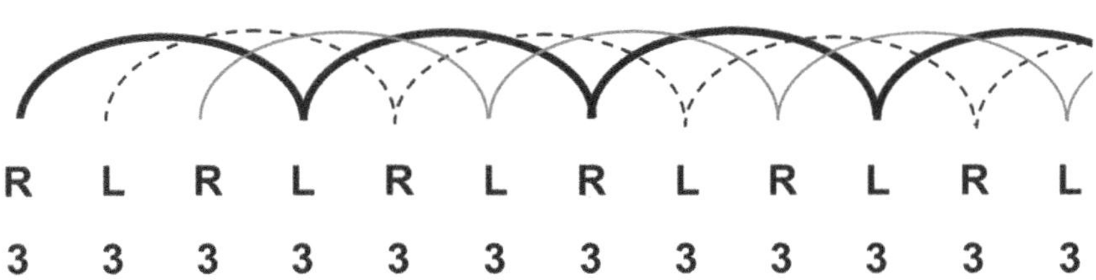

Figure 3.2 A space-time diagram for the (3) pattern.

Now that you know what a space-time diagram is for a 3-ball cascade, practice juggling the 3-ball cascade with intention. Be mindful of the beat and path of each ball.

Analyzing space-time diagrams for 3-ball juggling

Space-time diagrams are useful for depicting juggling patterns. Although all space-time diagrams of juggling patterns involve mathematical sequences (lists of numbers where order matters), not all mathematical sequences denote actual juggling patterns. In fact, we can study the space-time diagrams to determine whether or not a particular number pattern can actually be juggled. Recall that when using siteswap notation, we are assuming that the hands alternate, there is an even tempo, and all balls that arrive in a hand must leave that hand immediately.

Let's see if space-time diagrams can help us determine what juggling patterns are possible. This time, *you* will be creating the space-time diagrams.

8) Copy the diagram in Figure 3.3 and add arcs to see if (5, 4, 6) is a possible juggling pattern:

R	L	R	L	R	L	R	L	R	L	R	L
5	4	6	5	4	6	5	4	6	5	4	6

Figure 3.3 A space-time diagram for a (5, 4, 6) pattern.

As you trace the path of each ball, determine whether or not this could be a juggling pattern under the assumptions for siteswap notation we noted previously. How does tracing the path on your space-time diagram help you defend your claim?

9) Copy the diagram in Figure 3.4 and add arcs to see if (4, 3, 2) is a possible juggling pattern:

R	L	R	L	R	L	R	L	R	L	R	L
4	3	2	4	3	2	4	3	2	4	3	2

Figure 3.4 A space-time diagram for a (4, 3, 2) pattern.

10) Is (4, 3, 2) a possible juggling pattern? Why or why not?
11) You probably noticed that the previous sequences can't be juggled. Give the siteswap notation of another sequence that is impossible to juggle under our assumptions. What can you generalize from these examples?
12) How can you use siteswap notation (of just the beats) or a space-time diagram (indicating the balls' paths along with the beats) to determine when a juggling pattern is not possible?
13) We have examined some patterns that are impossible to juggle because the balls 'crash' or land at the same time in one hand. Draw space-time diagrams of juggling patterns that are impossible for a different reason.

IV. What else?

Different 3-ball patterns

Let's explore some other patterns and see what their respective space-time diagrams can tell us about their mathematical structures.

Watch video 3. In this segment, Sean Gandini demonstrates several different 3-ball patterns. If you are not familiar with juggling, it may be difficult to follow these patterns, but do your best to try to match the juggling with the space-time diagrams. You can always slow down the playback speed of the video.

Choose one (or more) of the patterns. Rewatch just that segment and try to match the juggling to its corresponding space-time diagram.

The space-time diagram for one of the patterns for 3-ball juggling featured in the video is shown in Figure 3.5.

Figure 3.5 A space-time diagram for the 3-ball juggling pattern (1, 5).

Describe what is going on in Figure 3.5 and juggle three balls in this way if you can.

1) What do you think is the significance of the sum (6) of the two numbers (1, 5) that appears in the siteswap notation for this juggling pattern?

Juggling 4 balls

Look at the space-time diagram for the **4-ball fountain** in Figure 3.6.

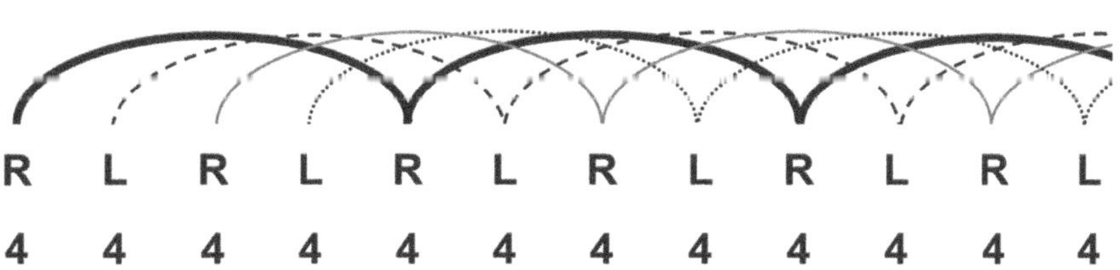

Figure 3.6 A space-time diagram for the 4–ball fountain (4).

Describe what is going on in Figure 3.6. If you can, juggle the pattern.

2) Describe two similarities and one difference between the 3-ball cascade and the 4-ball fountain.
3) What else do you notice about these juggling patterns?
4) Watch Video 4 to confirm or adjust your descriptions.

V. Story

Siteswap notation is one of a number of notations used for juggling. As you have seen, siteswap notation keeps track of the number of beats a ball is in the air. It is a notation that uses a sequence of numbers to represent juggling. The space-time diagram incorporates siteswap notation and tracks the path of the balls over time. Other representations of juggling, such as the ladder diagram (see Figure 3.7), also track the balls through time but in a different way. We will try to make sense of the ladder diagram in the 'Going beyond' section.

These different notations showcase creativity in mathematics. They represent innovative ways of keeping track of movement that happens in our 3-dimensional world over time in the form of 2-dimensional, static diagrams.

According to the *Oxford English Dictionary*, play is defined as an "activity engaged in for enjoyment and recreation, especially by children" (Oxford English Dictionary, n.d.). But adults play too! Juggling is just one example that connects mathematics with play. Perhaps you play games. When you play a video game, for example, you are undoubtedly strategizing and using your spatial visualization skills. Likewise, when playing a board or card game, you are often making decisions based on your understanding of probability.

Although a select few participate in sports in their line of work, most of us consider sports to be in the realm of play. In sports, there are many examples of mathematics in movement. For instance, in soccer, the players are constantly adjusting their speed, angle of shot, and direction, even if purely by instinct. Statistics also plays a vital role in sports, perhaps most notably in baseball. We typically think of statistics in baseball in terms of a description of a team or player's current standing and past accomplishments – something to know as a viewer or fan. But statistics can also be used by the players as the basis for strategies such as what pitch to throw or whether or not to steal a base.

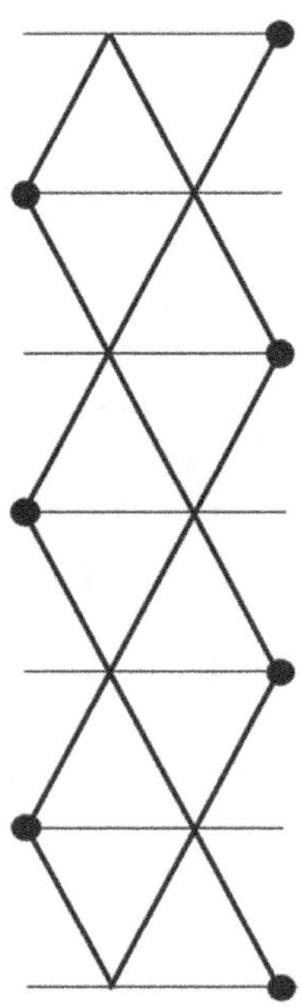

Figure 3.7 An example of a ladder diagram used for tracking juggling patterns.

Figure 3.8 Undergraduate mathematics students playing with mathematics. Photographs by Mike Peters – Montclair State University.

We also often associate play with toys. Think of the toys you grew up with. Can you connect any to mathematics? Where is the mathematics in jumping rope? Or flying a kite? Or playing with a yo-yo? Or building using Legos®? There are many aspects of these toys that can be analyzed mathematically – from movement, to location in space, to timing. When you think about toys from a mathematics perspective, it may make you appreciate them in a new way. The photos in Figure 3.8 show some undergraduate mathematics students 'playing' with mathematics. Can you see evidence of mathematics in what they are doing?

VI. Going beyond

1) Find a video on juggling that you find especially creative, inspiring, or artistic. It might look very different from the juggling patterns we've explored here.
   a) What appeals to you about the juggling in the video?
   b) What about the juggling patterns do you find creative?
   c) What is mathematical about this particular kind of juggling?
2) Complete a mini-research project on the mathematics of juggling. Who has studied it? What aspects of juggling have people studied?
3) Study space-time diagrams for more complex juggling patterns. Analyze the notation by describing the path of each ball.
4) Make connections between juggling and modular arithmetic (see Quest 4.1 for a preview of modular arithmetic). Where does modular arithmetic show up in juggling?

5) Can you imagine juggling more than 5 balls? What is the record for the number of balls juggled? Draw space-time diagrams for the 5-ball juggling patterns you find or imagine.
6) Figure 3.9 shows a pattern for 2-ball juggling. Using this diagram, describe what is going on. What is the left hand doing? What is the right hand doing? Remember that the solid and dashed lines indicate different balls. If you can, juggle the pattern.

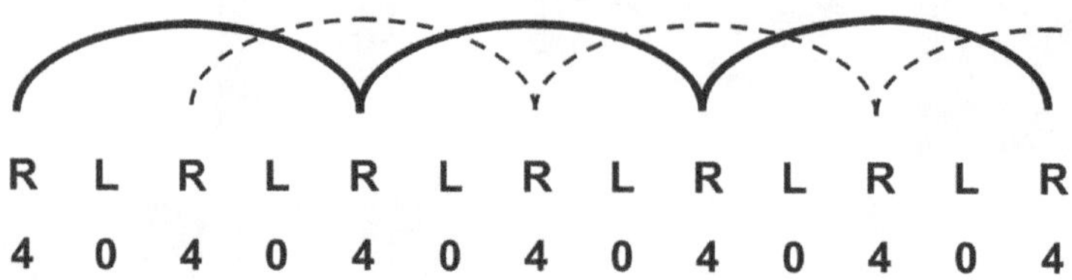

Figure 3.9 A space-time diagram for the 2-ball juggling pattern (4, 0).

7) Let's study ladder notation. Take a look at Figure 3.10.

   a) The letter 't' stands for time. What do you think the 'L' and the 'R' stand for?
   b) How many balls do you think are being thrown? Why?
   c) Can you juggle this pattern? (We bet you can!)
   d) What do you think is an appropriate unit for 't'?

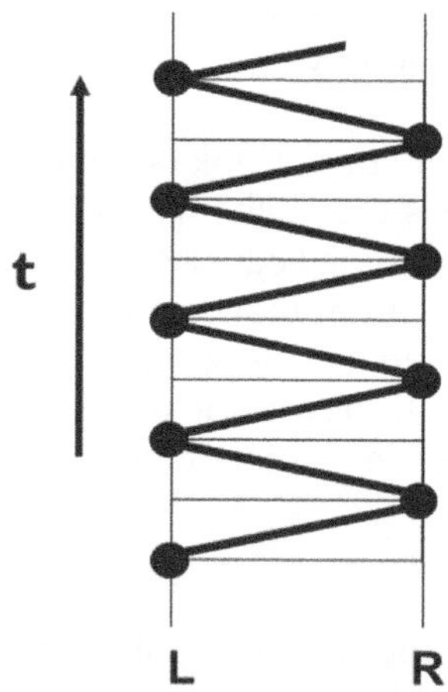

Figure 3.10 The ladder diagram to be used for Question 7.

8) Now that you have some ideas for how to read a ladder diagram, answer the following questions about Figure 3.11. Note that this represents the same juggling pattern as in Figure 3.7. We added some details, such as line style, to help you follow each ball.

   a) Describe the juggling pattern represented in Figure 3.11. How does this juggling pattern differ from the one represented in Figure 3.10?
   b) How many balls are being thrown?
   c) What is this juggling pattern called?

9) How are ladder diagrams different from space-time diagrams? Which do you prefer and why?

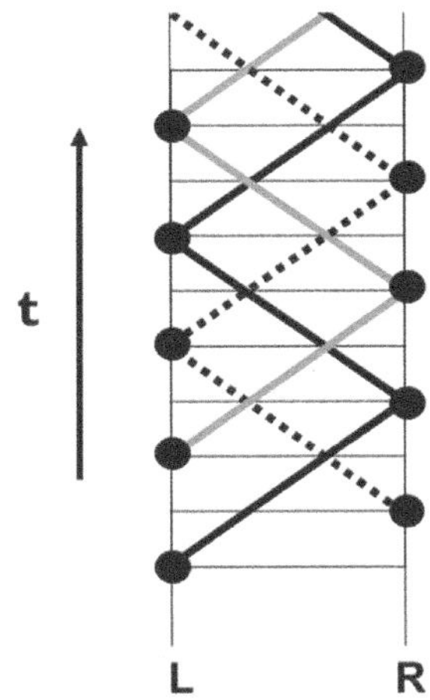

Figure 3.11 The ladder diagram to be used for Question 8 and 9.

VII. Let's reflect

1) Record a video of yourself juggling and use it as a tool to teach others about the mathematics of juggling. Include words such as beat, rhythm, pattern, and assumptions.
2) Juggling is often used as a metaphor for other phenomena. Write about what in your life you might be 'juggling'. Can you connect that phenomenon to mathematics? Can you create a mathematics notation that would help you track the 'juggling' that happens in your life? Could that notation help you 'juggle' more effectively?
3) If mathematics can be used to make sense of juggling, what other activities can be modeled with mathematics?

    a) Think about musical or artistic performances (for example, dance pieces) or childhood activities (for example, jump-roping) that you'd like to investigate. Generate a list of questions you might be able to answer using mathematics.
    b) Think about how you might use mathematical notation to represent the activity. Write out your notation.
    c) Find studies that have connected mathematics to your activity. Summarize their findings.

4) Write about the ways in which you were challenged in this quest, paying particular attention to the ways in which you were creative or used mathematical reasoning in new ways.

Quest 3.2: Playing with probability using a beachball and finger paint

Let's throw some math around!

In this quest, we are going to continue to 'play' as we engage with mathematics. You will count paint splatters as you toss a beachball with your fingertips dripping with finger paint. Things might get a bit messy, so keep those paper towels handy!

I. Invitation

Think of the last time you played with a beachball – maybe as a child or perhaps recently on a family vacation or in the park with friends. Tossing a beachball can be fun and relaxing, but who would have thought that there'd be a math lesson in it? An inflated beachball is shaped like a sphere, which might make you think of mathematics, but what about the act of tossing a beachball is related to mathematics?

II. What mathematics do you see?

For this activity, you will need a beachball with a pattern and a cup of washable finger paint. If you don't want to end up with colorful fingers, we suggest that you wear disposable gloves. Be sure not to wear your favorite white shirt!

With a partner, go outside and start tossing a beachball back and forth using both hands but only your fingertips (instead of your palms) and no paint (yet!) Continue this activity for a few minutes or until you are able to complete multiple tosses without dropping the ball.

Now that you've gotten the hang of tossing a beachball, dip your fingertips in paint. Have your partner do the same.

Now, carefully pick up the beachball with only your fingers or have a third person hand it to you. Begin tossing the ball back and forth with your partner, just like you practiced earlier. Watch the number of paint splatters grow with each new toss. Continue to toss the ball back and forth until you make at least 120 paint splatters. The more, the better. You can determine how many tosses to aim for beforehand – perhaps a 'back of the envelope' type calculation – or toss the ball back and forth for a few minutes until you think you have about 120 splatters. You may need to 'reload' paint on your fingertips from time to time.

When finished, carefully put the ball down and let the paint dry. You can wash your hands now!

While the paint is drying, answer the following questions:

1) What is mathematical about tossing a beachball with painted fingertips?
2) What mathematical problems could the paint splatters help you solve? Pose at least one mathematical question.
3) As you consider the paint splatters on the ball, make a conjecture related to the problem you posed.
4) What might you do to confirm your conjecture or answer your question?

III. Let's see what happens

There are many interesting mathematical questions we can pursue using a beachball, especially when it is being tossed. We are going to exercise our probabilistic reasoning by connecting mathematics to the tossing of a beachball.

Analyzing the splatters on your beachball

Count the number of splatters, which we will call dots, in each region of your beachball. For this exercise, we are assuming that your beachball has four different shadings. Because our book is in black and white, we have labeled these regions by color (see the yellow, blue, red, and white regions in Figure 3.12). We will base our questions on the beachball shown in Figure 3.12, but the one you tossed might look completely different – which is okay.

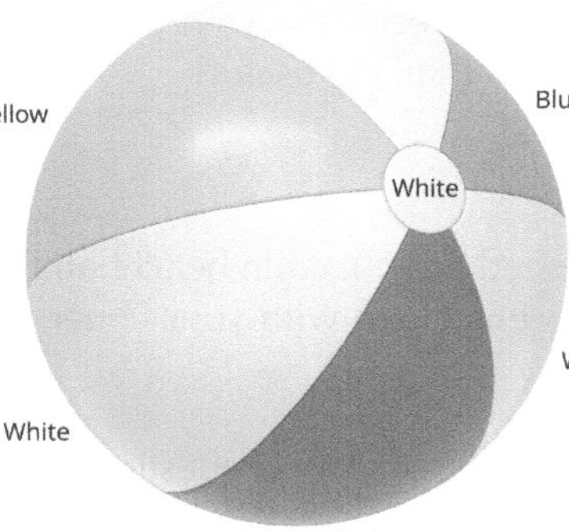

Figure 3.12 A beachball with blue, red, yellow, and white regions labeled. Image retrieved from Shutterstock and modified by authors.

Copy Table 3.1 and fill in the empty cells. Adjust the table to represent the patterns on your beachball as needed.

Table 3.1 A table for recording the distribution of dots on the beachball.

Regions	*Number of dots landing in the region*	$\frac{\textbf{Number of dots in the region}}{\textbf{total number of dots}}$ expressed as a percentage
Yellow region		
Blue region		
Red region		
White regions		
Entire beachball		100%

1) What mathematical questions come to mind as you view your table?

For the beachball shown in Figure 3.12, you might have wondered what region would have the greatest number of dots, or if the blue region and yellow region would have comparable numbers of dots. Perhaps you wondered if the number of dots in each region could be predicted from the pattern of the beachball. Let's see how your results align with what your predictions might have been.

2) What percentage of the dots would you have expected to land in each region (blue, red, yellow, and white)?
3) Does your data from tossing your beachball align with your predicted values? Explain your response.
4) What would have made your data align more closely with your predictions?
5) Do you think that your results could be replicated with a smaller beachball with the same pattern? A larger beachball?
6) What are some connections between this activity and probability?
7) In what contexts is this kind of thinking useful?

Tossing a globe

For this next activity, you will need an inflatable beachball that resembles a globe (see Figure 3.13). You are going to toss this globe with your fingers dipped in paint, as you did in the previous activity.

Before tossing the globe, consider the following questions and estimate each answer.

8) What is the percentage of Earth's surface that is covered by water?
9) What is the percentage of Earth's surface covered by the continent you live on?
10) How big is Antarctica, expressed as a percentage of Earth's surface? How big is it expressed as a percentage of the surface of Australia?

Figure 3.13 An inflatable globe.

Now toss the globe using washable paint on your fingertips as before to mark dots. Make an appropriate table to help you organize your data and refine your estimates to Questions 8–10.

11) Look up the actual answers to Questions 8–10 on the Internet.
12) How accurate were the results you obtained from this exercise?
13) What errors or discrepancies did you notice?
14) What could have led to these discrepancies?

IV. What else?

Tossing a beachball and counting paint splatters on its surface is a simplified example of a **Monte Carlo** simulation. Each paint splatter (or dot–where your finger touched the ball) is a data point that resulted from a random event (where the ball landed – where your fingertip touched the ball). In our case, we used our results to confirm a value that was known (the percentage of the beachball that was yellow, for example). In other cases, results can be used to determine the likelihood of a future event (for example, what chance is there that traffic will be bad at 7 p.m. tonight?).

The use of this kind of simulation is called the Monte Carlo method, where we use random sampling to estimate the likelihood of various outcomes. For example, when rolling dice, we know that the probability of rolling a '5' on a single fair six-sided die is ⅙. We can also estimate this probability using the Monte Carlo method, by rolling a single fair die many times (let's say 10,000 times) to see what fraction of the rolls lands on '5'.

There are many applications of this method. We will continue to use it to estimate how the sizes of different regions of the world compare to each other. Next, we are going to see what happens to our estimates when we represent the world on a flat surface.

Considering areas on a map

Find a poster-size paper map of your state, province, or analogous geopolitical region. Scatter 'dots' onto your large map by placing it on the ground when it's raining (but not raining too hard!). Alternatively, scatter dots with finger paint onto the map in a haphazard manner. If you do not have access to a poster-size paper map, you can use a map on your computer. If you are working with an electronic image, think of how you can program the random scattering of 'dots'.

1) Use methods developed in this quest and tools of your choice to determine the portion of the area of your region that is covered by forest.
2) Next, see if you can determine how much of the area of your region is covered by water.
3) Describe the calculations you used to make your estimates.
4) Think of other questions related to your map that you can answer using the Monte Carlo method. Describe how you would answer each question.

Considering the game of darts

If you can, get a hold of a dart board. If you don't have a dart board handy, that is okay. Make your own version, adapting the questions that follow accordingly, and use a suitable substitute for darts. (This is where you can be really creative.) Now try throwing darts haphazardly (without aiming) at your dart board (see Figure 3.14). It's okay if some hit the wall or land on the floor.

Figure 3.14 A dart board.

Photograph by Eric Mclean, retrieved from https://www.pexels.com/photo/worn-dart-board-on-green-wall-9899849/

5) What questions can be answered through a Monte Carlo simulation with a dart board? Describe how you would set up and conduct a Monte Carlo simulation to answer your questions.

Collect data for a large number of haphazard throws (without aiming) and note where each dart lands. Note that the result of a throw (where the dart lands) is analogous to each dot we tracked in the beachball simulation. We leave it to you to see how many throws you think would suffice.

6) Use your data to approximate the likelihood of hitting the bull's eye (the smallest circle at the center of the board if you are not aiming).
   a) What percentage of your throws landed in the bull's eye? (Make a special note of which throws you are counting in your calculations.)
   b) Calculate the actual percentage of the dart board taken up by the bull's eye by other means.
   c) How do these compare?
7) Use your data to approximate the likelihood of hitting the slice of the board represented by '20'. Let's assume that a slice includes all scoring areas that are contained within it but not the bull's eye.
   a) What percentage of your throws landed in the '20' slice?
   b) Calculate the actual percentage of the dart board taken up by the '20' slice by other means.
   c) How do these compare?

8) What if you aim this time and *try* to hit a specific slice on the dart board? How can this affect the possible outcome? Note that in Questions 5–7, we were asking about a situation where the darts are being thrown haphazardly. In this question, you are aiming at a particular slice.
9) How did you account for the throws that did not land on the dart board in your responses to Questions 5–7?
10) Pose your own question related to the dart board and answer it.

V. Story

Monte Carlo simulations give us information that allows us to create mathematical models about phenomena that are difficult to analyze. It relies on repeated random sampling, often using computer algorithms. In this quest, the point where each finger touched the ball (each paint splatter) was one data point. With each toss, you actually collected 10 data points that helped you estimate the relative area of each section of the beachball.

The Monte Carlo method is frequently used in physics, biology, chemistry, psychology, business, artificial intelligence, and computer science as well as sociology, political science, and finance. For example, when attempting to predict how profitable a financial portfolio will be, analysts run Monte Carlo simulations to determine possible outcomes of the portfolio using past trends and patterns of the market. From there, they can help clients decide how to allocate funds and adjust their portfolio based on these risk assessments.

Let's look at the weather. Have you noticed how weather predictions are often wrong? Weather scientists used to think that if they could just measure today's weather closely enough and develop good enough models, they could predict the weather. It turns out that by its very nature, weather is **chaotic**. A tiny imprecision in current weather measurement can lead to wildly incorrect predictions of future weather. So, weather forecasters have turned to Monte Carlo simulations to attempt to get better results. They run simulations of past weather patterns under various conditions to come up with their weather predictions. Consider other professions that involve attempting to forecast the future. In what ways might Monte Carlo simulations be useful to those fields?

VI. Going beyond

1) Write a computer code that utilizes the Monte Carlo method. For example, you can use the random number generator feature on a spreadsheet to generate your random samples. You might start with a simple example: If you randomly generate 10 single digit numbers, what percentage of the time do you get a number less than 3? With this practice, you can consider more sophisticated examples for your design.
2) Research the use of the Monte Carlo method in meteorology. What did you find the most interesting? What surprised you? What made you wonder or want to know more?
3) How can the Monte Carlo method be used in the following fields?

    a) Sports
    b) Insurance
    c) Biology
    d) Physics

4) Propose how one might use the Monte Carlo method in a field not listed in the previous question.

5) Imagine that you've been saving to buy a ticket to see your favorite baseball player. Very little would make you happier than to catch a home run ball hit by this player. You are selecting your seats. What data would be helpful to you in making this decision? Collect that data and come to a determination on where the best seat would be (see Figure 3.15).

Figure 3.15 A baseball field. Photograph by Pixabay, retrieved from https://www.pexels.com/photo/aerial-view-of-baseball-field-264279/.

6) In what ways can a Monte Carlo simulation be useful in each of the following situations?
   a) Making a financial decision (for example, whether to buy a new or used car)
   b) Deciding the mode of travel or route for commuting to work or school
   c) Deciding on which day of the week to buy an airline ticket for the best price
7) Research the history of events that have been modeled using the Monte Carlo method. Make a timeline of key events, individuals, and outcomes associated with this method. What are your major takeaways? Did anything surprise you?

VII. Let's reflect

1) What are the advantages of the Monte Carlo method for estimating outcomes or numerical results? What are the disadvantages of the Monte Carlo method for estimating such values?
2) What surprised you about the Monte Carlo method in this quest? What stood out to you as being especially significant about this method?
3) How did you tap into creativity to complete this quest? What elements of the Monte Carlo method required a creative approach?
4) In what ways is the Monte Carlo method limited? How can we adjust for these limitations?
5) When in your life have you relied on past experiences to make decisions? How similar (or not) was that process of decision making to Monte Carlo simulations?
6) How did learning about the Monte Carlo method help you appreciate variability or uncertainty?

Quest 3.3: A different take on π with gummy bears and Monte Carlo simulations

Where's the π?

In the previous quest, you learned about the Monte Carlo method. You explored simulations based on random outcomes, like where your fingers touched a beachball as you were tossing it. As a result, you were able to compare the areas of the different sections of the beachball without using geometric formulas. You also saw how Monte Carlo simulations are used to determine the likelihood of events occurring. Now let's see how we can use the Monte Carlo method to estimate the famous mathematical number pi (π). We are going to do this using techniques similar to those highlighted in the previous quest, combined with slightly different tools and mathematical methods. Let's get started!

I. Invitation

Pi is one of the most famous and thought-provoking numbers known in mathematics, and without question, π deserves every bit of the intrigue it has garnered in the past four thousand years.

1) Do you remember learning about π in geometry class? Do you recall seeing π when studying the set of **real numbers**? If yes, what do you remember about it?

In this quest, you will estimate this special number using the Monte Carlo method. But before we get started, take a look at the first 31 digits of π listed here. The three dots at the end indicate that the digits of π continue forever.

3.141592653589793238462643383279 . . .

2) What do you notice about π?

Perhaps you learned previously that π is an **irrational number** – one that cannot be expressed as a ratio of two **integers** – or that π is the ratio of the **circumference** of a circle to its **diameter**. In this quest, you are going to consider π from a different perspective: How can the Monte Carlo method help you estimate the value of π?

II. What mathematics do you see?

Look at Figure 3.16. The outer boundary in this figure is a square.

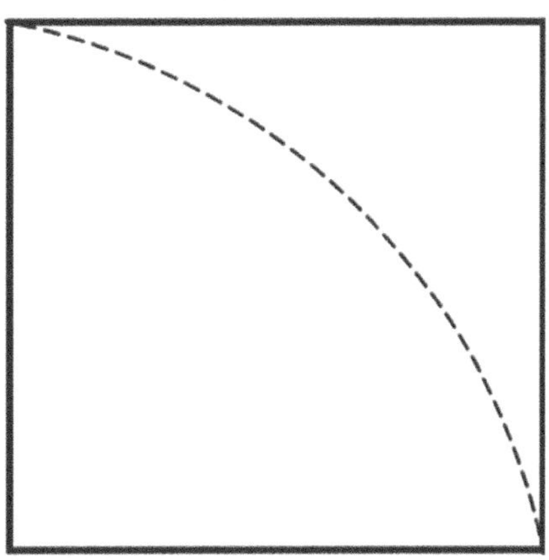

Figure 3.16 A quarter circle inscribed in a square.

In what ways is π represented in Figure 3.16?

Let's say that the length of each side of the square is 1 meter. What can you say about the relationship between the area inside the dotted region and the whole square?

Recall that π is the ratio of the circumference, C, of a circle to its diameter, d: $\pi = C/d$. Pi is also used to determine the area, A, of a circle: $A = \pi r^2$, where r represents the radius of the circle. We will use this information later.

III. Let's see what happens

Our goal is to use the Monte Carlo method to estimate the value of π. Use chalk to draw a 1 meter by 1 meter square on a flat surface or create it using masking tape. You can use a flat sidewalk or space on a playground. Alternatively, you can draw your square on a big piece of poster paper. (You might need to tape poster paper sheets together to make a larger piece.)

Draw a quarter circle inside the square so that the boundary of your quarter circle meets opposite corners of the square (see Figure 3.16). It might help to imagine the whole circle – the center of it should be at a corner of the square. You can attach a pen (or chalk, if you are outside) to a piece of string and use the fact that all points on a circle are equidistant from the center to help you do this. Note that you are using the string and pen in a creative way!

Take about five handfuls of gummy bears (small pieces of candy) and sprinkle them haphazardly inside the square from about 10 centimeters above the square. You can also use dried beans, coins of the same size, or pieces of a different type of small candy. They should be somewhat evenly distributed within the square. Consider just the ones that land in the square.

1) How can you use the distribution of gummy bears to estimate the value of π? Draw from what you learned in the previous quest (Quest 3.2).
2) Estimate π using the distribution of gummy bears in your square. What is your estimate?

Some hints

Feeling stuck? That's okay! You can either step away and come back to this later or answer the following questions meant to guide you. (Skip to Question 8 if you're ready to explore further.)

3) What fraction of the gummy bears in the square landed within the quarter circle?
4) What is the significance of this fraction?
5) What is the calculated area of the square based on the length of the sides?
6) What is the calculated area of your quarter circle in terms of π?
7) How can you use this information to estimate the value of π?

Repeating the experiment

8) Repeat this experiment at least five times and record your results in a systematic way. Use these results to determine an average estimated value of π. How does this

estimate compare to your initial estimate of π? How does it compare to the actual value of π?

9) If possible, collect class data and use it to determine a new estimate. How does it compare to your initial estimate of π? How does it compare to the actual value of π?
10) Besides repeating the experiment on your own or combining your data with data collected by your classmates, what else would help you improve your estimate of π?
11) Write a paragraph about the process of using the Monte Carlo method to estimate the value of π in this way.

IV. What else?

To further test your ideas regarding the power and limitations of the Monte Carlo method, draw a new square, this time making it 0.5 meters by 0.5 meters. Inscribe a whole circle in the square. (You can use your chalk and string again!) Make sure that the circle and the square only meet at the four midpoints of the square's sides (see Figure 3.17). You are going to repeat the experiment from the 'Let's see what happens' section with this new setup. You will be scattering the same amount (five handfuls) of gummy bears and using your data to estimate π.

1) How do you think your estimate will differ from the estimate you obtained from the quarter circle experiment?

Now repeat the Monte Carlo simulation using about the same number of gummy bears as before (five handfuls) and the entire square and circle (see Figure 3.17) and record your results.

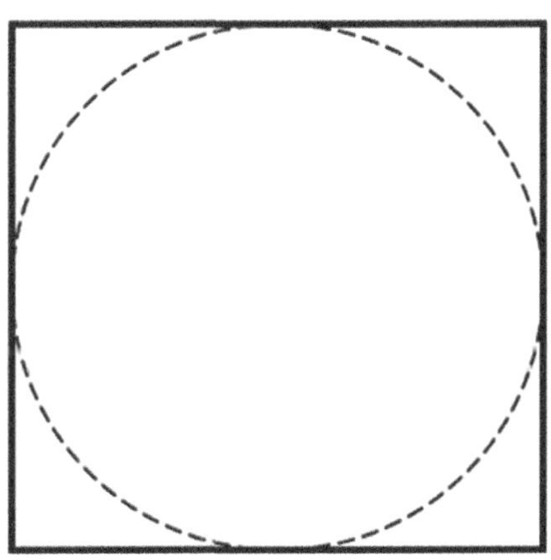

Figure 3.17 A circle inscribed in a 0.5 meter by 0.5 meter square.

2) Use your data to estimate the value of π.
3) How did your results compare to those obtained earlier using the quarter circle and original (larger) square?

4) How did changing the radius and the portion of the circle (quarter to whole) inscribed in the square affect your calculation? How did keeping the number of gummy bears the same impact your estimate of π?
5) In what other ways could your estimate of π be improved?

V. Story

Pi represents the ratio of a circle's circumference to its diameter. This proportional relationship holds for all circles, regardless of their size. If you need some more convincing that this is true, use a paper ruler or measuring tape to find the circumference and diameter of circles you see around you (a coffee cup, a coaster, a bicycle tire, a cooking pan). Find the ratio of each circumference to its diameter and express each in decimal notation. What did you find?

Pi is an irrational real number. An irrational number cannot be represented by a ratio of integers, and its decimal representation consists of infinitely many digits with no repeating numerical pattern. For comparison, 3.141514151415 . . . is close to π but rational because it can be written as a mixed number where the fractional part is a quotient of two integers: $3 + \frac{1415}{9999}$. Irrational numbers do not contain these types of repeating patterns. Here are the first 31 digits of π again:

3.14159265358979323846264338327 9 . . .

Many ancient civilizations, including the Babylonians, Egyptians, and Greeks, determined their own approximations of π. For example, the Babylonians (circa 1900–1680 BC) estimated π to be 3.125, but sometimes used 3 out of convenience, while the ancient Egyptians estimated π to be approximately 3.16, as shown in *Rhind Papyrus* (circa 1650 BC). In Greece, Archimedes of Syracuse (287–212 BC) inscribed and circumscribed polygons within and about a circle to give a lower and upper bound to the area of the circle and estimated π to be between 3.1408 and 3.1429 (see Figure 3.18).

People's fascination with π is just as strong today as it was almost 4,000 years ago. We can use computers to calculate additional digits of π and have found trillions of them. The race to find the next digit of π continues! People also enjoy competing with each other over who can memorize more digits of π. Many people even celebrate π day on March 14 every year. Can you guess why?

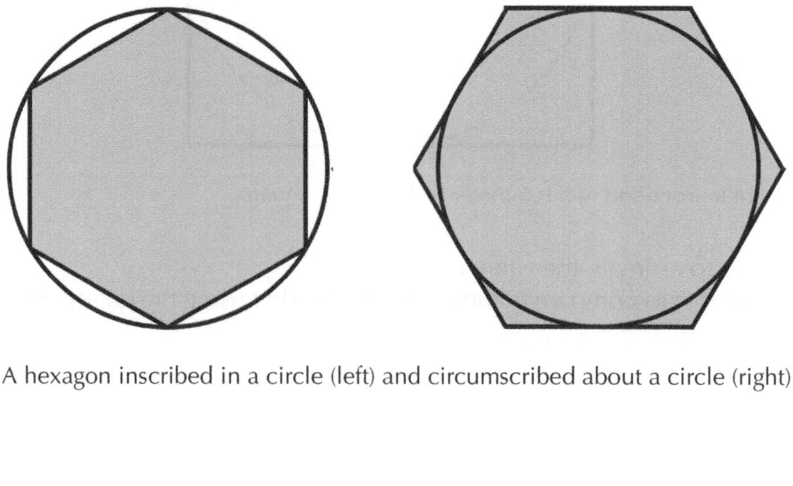

Figure 3.18 A hexagon inscribed in a circle (left) and circumscribed about a circle (right).

VI. Going beyond

1) Wisława Szymborska, a Polish poet, essayist, and winner of the 1996 Nobel Prize in Literature, wrote a poem *Pi* (Szymborska, 1995). Read this poem. What is the author's interpretation of π? What new understanding of π did you gain from this poem?
2) Write your own poem about π.
3) Research George Cantor's work on the size of infinite sets. What did Cantor suggest about the size of the set of rational numbers? The irrational numbers? The real numbers? Did any of his findings surprise you? Why? Make a diagram showing your interpretation of the relationship between these sets.
4) Research the history of π, especially the work completed after Archimedes' famous estimate. Make a timeline of major discoveries related to π.
5) Do research on the method for approximating π using Pick's theorem. Compare this method to the Monte Carlo method.
6) Suggest other shapes whose areas can be estimated using the Monte Carlo method. How would you go about estimating these areas?

VII. Let's reflect

1) Which part of the Monte Carlo method for estimating π did you find the most challenging? Why?
2) Has your understanding of π changed in any way? How did learning about different ways to estimate the value of π change your thinking about this irrational number?
3) Make a video teaching someone how to estimate π using the Monte Carlo method.
4) What mistakes did you make in this quest? What challenges did you face? How did you overcome them?

Project 3: Math book club

Reading about mathematics? Of course! Believe it or not, many books for the general audience – both fiction and nonfiction – have been written about mathematics. As shown in this chapter, physical activities like juggling can offer opportunities to engage with mathematics through play. On the other hand, play can be a mindset that we adopt when faced with a mathematical problem. For this project, you will read a book that showcases an element of play in mathematics in one way or another. The book you choose may describe the curiosity, exploration, imagination, and improvisation – characteristics also associated with play – that led to a famous mathematical discovery. Or, the book might feature an activity associated with play from a mathematical perspective. Either way, you will learn about creativity in mathematics as presented through a book written for a wide audience.

Just as a group of friends might gather to discuss the current number one bestseller, you will be forming a book club with your friends or classmates to discuss a book about mathematics.

Although for the purposes of the project you will be discussing a single book, the purpose of a mathematics book club is to expose you to fascinating and accessible books about mathematics and to bring your attention to the different ways in which we can experience mathematics. Through these books, you could learn that there is much more to mathematics than what you've experienced in the past.

The objectives of the book club are to

- Provide a safe, cozy, and friendly environment for you to chat about a book that relates to mathematics and creativity.
- Encourage you to think about how the themes in the book connect to creativity and mathematics, especially with respect to play.
- Deepen your appreciation for mathematics as a discipline full of surprises, play, strategies, and fascinating stories.
- Exemplify how reading, literature, discussions, and mathematics can go hand in hand.
- Have you explore a mathematics topic that interests you.
- Provide an enjoyable reading activity.

Contact your friends, classmates, or family to form a book club of about four people. Having your friends join you on this adventure will make the project more interesting and will provide a social environment in which you can discuss your insights. You can add snack duty, have it outdoors, meet online . . . introduce your own twist!

Here are some guidelines to help you get started.

Choose a book

To begin, choose a book with your friends. We offer some suggestions of books about mathematics and mathematicians that we think highlight the central role of curiosity, inquisitiveness, and play in mathematics:

1) *Mathematics for Human Flourishing* by Francis Su
2) *Einstein's Dreams* by Alan Lightman

3) *Chaos: Making a New Science* by James Gleick
4) *The Joy of Mathematics* by Theoni Pappas
5) *Fermat's Enigma* by Simon Singh
6) *Does God Play Dice? The New Mathematics of Chaos* by Ian Stewart

Set up a schedule

Determine a schedule for your book club. Plan about four hour-long meetings. Depending on the length of the book, you might choose to have biweekly or monthly meetings. Assign chapters to be discussed at each of those meetings, allowing ample time to read the chapters between meetings.

Set group norms

Consider setting group norms. These norms will help ensure that everyone contributes to the discussion and feels comfortable sharing their thinking. The norms can be about how to structure each meeting (for example, have questions sent out two days ahead of time) or about group behaviors (for example, try not to interrupt others). These norms can be read aloud before each meeting as a reminder.

Assign roles

In addition to setting group norms, having designated roles for each person is another way to get everyone involved. Assign roles so that each book club member is host and scribe at least once. The following are our suggestions for each of the assigned roles. Of course, if your group (or instructor) decides to keep it more free-form to encourage spontaneity, we welcome that as well.

Host

The host runs the meeting. They are responsible for preparing discussion questions and sharing them with the group before their session. Of course, other members of the club can also share questions for discussion or bring in artifacts (for example, a drawing or a poem). The host is also responsible for starting the conversation and making sure that it continues during the meeting.

There is no suggested number of discussion questions, but there should be enough to hold a conversation that will last at least an hour. (Better to have too many questions than too few.) The host should assume that the other participants have read the section of the book being discussed. The questions should not be meant to test the reading comprehension of the other participants but rather provide a learning and thinking space to enjoy mathematics through meaningful discussions. The questions should be interesting, thought-provoking, and when possible, draw out the connections between mathematics, creativity, and play. The host might choose to bring in pictures of artwork, historical essays, articles, and other resources that might add to the discussions. For an added dimension, the host can prepare materials such that the group can do mathematics together, inspired by a task or story in the book.

We do encourage impromptu conversations as well, so use your prepared questions thoughtfully. It may be that you only get to a few of them because the conversation took another turn. That's totally fine!

Scribe

The scribe takes notes at each meeting. They are responsible for ensuring that the main discussion points, ideas, and reflections at each meeting are recorded and organized for future reference. As the group gets further into the book, the conversations will likely deepen with each subsequent meeting and discussion. The notes that the scribe takes will be shared with the group and will help the members see how far they've come. The scribe can also close the meeting with a brief 2–3-minute summary of the key takeaways and pose a reflection prompt for everyone to think about for the next meeting.

Again, your group can decide how structured you would like your book club to be. Depending on your situation, you might decide that having no roles or reducing the responsibilities of each role allows for more free-form discussions.

We hope this project exposes you to books that you had not considered picking up before. As your knowledge and perspectives about mathematics (and other disciplines) broaden, consider delving deeper into topics of interest to you by reading trade books including fiction, biographies, and historical books about mathematics written for a wide readership. So let this project be a springboard for you to continue combining your academic pursuits with leisurely reading!

Chapter introduction: Cited and related works

Burghardt, G. M. (2011). Defining and recognizing play. In A. D. Pellegrini (Ed.), *The Oxford handbook of the development of play* (pp. 9–18). Oxford University Press.
Ervynck, G. (1991). Mathematical creativity. In D. Tall (Ed.), *Advanced mathematical thinking* (pp. 42–53). Kluwer Academic Publishers. https://doi.org/10.1007/0-306-47203-1_3
Kaufman, J. C., & Beghetto, R. A. (2009). Beyond big and little: The Four C model of creativity. *Review of General Psychology, 13*(1), 1–12. https://doi.org/10.1037/a0013688
Oxford English Dictionary. (n.d.). *Play*. In *OED online*. Oxford University Press. Retrieved June 26, 2025, from https://www.oed.com/
Torrance, E. P. (1966). Nurture of creative talents. *Theory Into Practice, 5*(4), 167–173. https://doi.org/10.1080/00405846609542020

Quest 3.1: Cited and related works

Buhler, J., & Graham, R. (1984). Fountains, showers, and cascades. *The Sciences*, 44–51.
Burghardt, G. M. (2011). Defining and recognizing play. In A. D. Pellegrini (Ed.), *The Oxford handbook of the development of play* (pp. 9–18). Oxford University Press.
Ervynck, G. (1991). Mathematical creativity. In D. Tall (Ed.), *Advanced mathematical thinking* (pp. 42–53). Kluwer Academic Publishers. https://doi.org/10.1007/0-306-47203-1_3
Monahan, C., Munakata, M., Vaidya, A., & Gandini, S. (2020). Inspiring mathematical creativity through juggling. *Journal of Humanistic Mathematics, 10*(2), 291–314. https://doi.org/10.5642/jhummath.202002.14
Naylor, M. (2012). The mathematical art of juggling: Using mathematics to predict, describe and create. In *Proceedings of Bridges 2012: Mathematics, Music, Art, Architecture, Culture* (pp. 33–40).
Oxford English Dictionary. (n.d.). *Play*. In *OED online*. Oxford University Press. Retrieved June 26, 2025, from https://www.oed.com/

Quest 3.2: Cited and related works

Monahan, C., Munakata, M., & Vaidya, A. (2020). Engaging in probabilistic thinking through play. *Mathematics Teacher: Learning and Teaching PK-12, 113*(9), 18–23. https://doi.org/10.5951/MTLT.2019.0198

Quest 3.3: Cited and related works

Szymborska, W. (1995). Pi. In S. Baranczak & C. Cavanagh (Trans.), *View with a grain of sand: Selected poems* (pp. 66–67). Harcourt Brace & Company.

Project 3: Cited and related works

Gleick, J. (1988). *Chaos: Making a new science*. Penguin Books.
Lightman, A. (1993). *Einstein's dreams*. Warner Books.
Munakata, M. (2005). Welcome to the math book club: A new approach to integrating mathematics, reading, and writing. *PRIMUS – Problems, Resources, and Issues in Mathematics Undergraduate Studies*, *15*(3), 259–266. https://doi.org/10.1080/10511970508984120
Pappas, T. (1993). *The joy of mathematics: Discovering mathematics all around you*. Wide World Publishing.
Singh, S. (1998). *Fermat's enigma: The epic quest to solve the world's greatest mathematical problem*. Anchor.
Stewart, I. (2002). *Does God play dice? The new mathematics of chaos* (2nd ed.). Wiley-Blackwell Publishing.
Su, F. E. (2020). *Mathematics for human flourishing*. Yale University Press.

4 Where is the beauty in mathematics?

When people think of creativity, they often think of the 'arts': for example, painting, sculpting, or composing music. The results of these activities are often also associated with beauty. We comment on the beauty of the sound of musical notes or the technique used to make a painting beautiful. It is probably safe to say that mathematics is not something that immediately comes to mind when most people think of 'art' or 'beauty'. But mathematics can be beautiful as well!

In what ways can mathematics bring our attention to the beauty around us? Also, is it possible for mathematics to inspire us, even surprise us, with beauty? These are some of the questions you will explore in this chapter. While our journey thus far has involved thinking creatively as we engage with mathematics, this chapter makes explicit connections between mathematics and aesthetics. By drawing out the beauty in our world through mathematics and seeing a different kind of beauty within the world of mathematics itself, our aim is to encourage you to appreciate the various ways that beauty can be experienced.

The creative process can often lead to aesthetically pleasing products (Cropley & Cropley, 2008). When we tap into our creativity when doing mathematics, the result can be a piece of art (as typically understood), or it can be a beautiful and elegant mathematical discovery. As we mentioned earlier in this book, we are hoping that you experience 'little-c', or 'little creativity' (Kaufman & Beghetto, 2009), where the discoveries are new to *you* even if they may have been previously discovered by others. These discoveries can include new connections, new applications of familiar topics, and new questions.

The quests in this chapter have you develop, examine, and admire mathematical ideas for their elegance and beauty. What is beautiful about numerical patterns and the connections between them? What makes a particular mathematical procedure artistic? What can be elegant about a mathematical process? You will also create art using mathematics as you think deeply about the mathematics that is driving that creation. You will use visual representations to revisit familiar topics (like square numbers) and view nature in a new light, through a mathematical lens.

DOI: 10.4324/9781003541097-5

Chapter 4: Where is the beauty in mathematics?

Quest 4.1: Making art with modular arithmetic

Give remainders a chance!

This quest introduces you to **modular arithmetic** – the mathematics where we are concerned with what *remains* when an integer is divided by an integer – and to beautiful visual representations of numerical patterns. You will explore modular congruences and relationships that result from adding and multiplying numbers in modular arithmetic. You will also design visual models of these relationships. You are about to see mathematics in a new light. It's time to connect the dots!

I. Invitation

A typical analog clock (see Figure 4.1) represents 12 hours. The one in our figure consists of two hands: a minute hand (the long hand) and an hour hand (the short hand). In this quest, we will concern ourselves only with the hour hand. We often differentiate morning from afternoon by saying either 5 a.m. (morning) or 5 p.m. (afternoon). But what happens when we consider a different system for referring to time?

Figure 4.1 A 12-hour analog clock. Drawing by Ashley Etheridge.

1) In military time, where the 24 hours of the day are represented, what hour does 16 o'clock (16:00) correspond to on a 12-hour clock? How about 23 o'clock (23:00)?
2) Now, straying away from the usual conventions of how we note time, what would be the equivalent of 42 o'clock on our 12-hour clock? What would 31 o'clock be?

We can use the same kind of thinking when considering the passing of time in hours.

3) If it is currently 4 o'clock, what time will it be 29 hours from now? How did you determine this?

Perhaps you doodled or jotted down quick calculations. Maybe you imagined or drew the hour hand circling around a clock. The hour hand 'wraps around' the clock a certain number of times, and what remains tells you the time. Whatever method you used in thinking about these questions, you were being asked to do calculations that involve modular arithmetic. Modular arithmetic is the arithmetic of cycles (wrapping around) and their resulting remainders. It is based on a famous theorem known as the **division theorem**. Even if you aren't familiar with this specific theorem, you have probably come across its results when you first learned division.

Let's say that we are dividing an integer a (the dividend) by an integer b (the divisor). The theorem states that a can be expressed as b times a unique integer quotient q plus a unique non-negative integer remainder r where the remainder r is less than the absolute value of b.

$$a = b \times q + r$$

Before going on, try to make sense of the division theorem.

4) What do you think the division theorem means? Try explaining the theorem to a classmate or stranger.
5) Why does the remainder have to be less than the absolute value of b, the divisor?

In our earlier examples, we were concerned with what remains after taking into account half-day (12 hour) or full-day (24 hour) cycles. We were thus interested in the remainder that results when the hour is divided by either 12 or 24.

Let's go back to Question 3. In that instance, 29 divided by either 12 or 24 results in a remainder of 5:

$$29 = 12 \times 2 + 5 \text{ and } 29 = 24 \times 1 + 5$$

Note that here, we are less concerned with how many times the divisor (12 or 24) divides 29 (twice or once, respectively). We are more interested in the remainder. We use the remainder of 5 to determine that if it is 4 o'clock now, in 29 hours it will be 9 o'clock because it would be five hours past 4 o'clock: $4 + 5 = 9$.

The following congruence statement shows that when you divide 29 by 12 or by 24, the remainder is the same:

$$29 \ (mod\ 12) \equiv 29 \ (mod\ 24)$$

This statement reads "29 *mod* 12 is congruent to 29 *mod* 24". This means that the remainder when dividing 29 by 12 is the same as the remainder when dividing 29 by 24. '*Mod*' is shorthand for **modulo**. Here we are using '*mod* 12' or '*mod* 24' to indicate division by 12 or 24. The symbol in the middle that looks something like an equals sign, but uses three horizontal lines instead of two, indicates **congruence**. These two expressions $29\ (mod\,12)$ and $29\ (mod\,24)$ are congruent because when you divide 29 by 12 or 24, the remainder is 5.

We can say that two integers are **congruent modulo *n*** when they have the same remainder when divided by *n*. For example, 14 and 38 are congruent modulo 12, written $14 \ (mod\ 12) \equiv 38 \ (mod\ 12)$, because they both have a remainder of 2 when divided by 12. Next is another useful way to make sense of this relationship.

$14 = 12 \times 1 + 2$
$38 = 12 \times 3 + 2$

6) Find three other numbers congruent to 38 and 14 in modulo 12. Think about different ways to approach this question.

II. What makes you wonder?

Imagine an analog clock that shows 8 hours instead of 12 hours. That is, this clock only has the numbers 1, 2, 3, 4, 5, 6, 7, and 8.

1) Draw an 8-hour analog clock.
2) What would 12 noon be on the 8-hour clock?
3) What would the equivalent of 17 o'clock be on the 8-hour clock?
4) If it is currently 5 o'clock, what time will it be 29 hours from now on an 8-hour clock? How did you determine this?
5) What assumptions did you make as you thought about these questions? How else could these questions be interpreted?

Numbers congruent modulo 8 share the same remainder when divided by 8. For example, 10 and 18 are congruent modulo 8 ($mod\ 8$) because they both have a remainder of 2 when divided by 8.

$10 \ (mod\ 8) \equiv 18 \ (mod\ 8)$

6) What numbers are congruent to 6 modulo 8? List at least two different numbers. Explain your reasoning.
7) Would you imagine that the hands of an 8-hour clock move faster, slower, or at the same speed as the hands of a 12-hour clock? State your reasoning. What assumptions are you making? How might someone else interpret the situation?

III. Let's see what happens

The mathematics of modular arithmetic can lead to beautiful patterns. Let's explore some of these patterns. We are going to consider the following expression: $kx\left(mod\ 8\right)$. This expression indicates that we will be working with modulo 8 (as in the second clock example). The k is a positive integer constant that we will modify for different examples and the x is our variable.

For example, consider the case where $k = 6$:

$6x\left(mod\ 8\right)$

Copy Table 4.1. The first column includes values of x. As you go across each row, the second column shows the values of $6x$. The values in the third column represent $6x\left(mod\ 8\right)$.

1) Calculate $6x$ for the given values of x and place each result in its respective row in the second column.
2) Then, calculate $6x\,(mod\,8)$ for the given values of x and place each result in its respective row in the third column.

One row (for $x = 2$) has been filled in as an example. Note that when $x = 2$, $6x = 12$, and the remainder when 12 is divided by 8 is 4. We can represent this as $12\ (mod\ 8) \equiv 4$.

3) Complete the table for all rows. Explain why the third column, $6x\,(mod\,8)$, contains only even integers. (Hint: look carefully at the second column.)

Table 4.1 A table for $6x\,(mod\,8)$.

x	$6x$	$6x\,(mod\,8)$
0		
1		
2	**12**	**4**
3		
4		
5		
6		
7		

Now let's see what happens when we connect our values for x with our values for $6x\,(mod\,8)$. For this, we are only concerned with the first and last columns of Table 4.1.

4) Draw a circle with eight evenly spaced points where each point represents a possible remainder when dividing by 8 (see Figure 4.2). This will resemble our 8-hour clock, with a 0 replacing the 8, since $0 \equiv 8\ (mod\ 8)$.

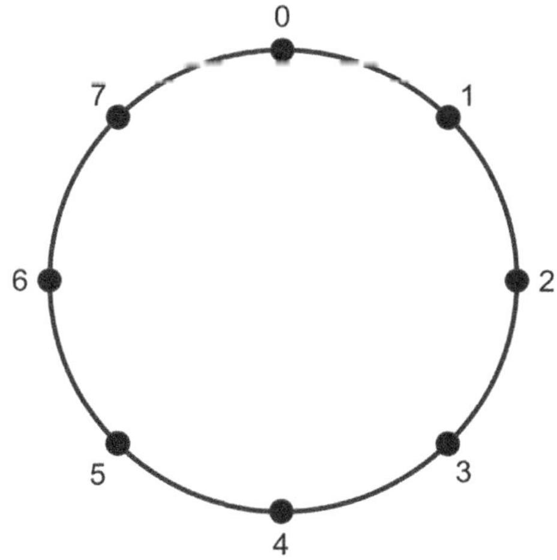

Figure 4.2 A circle with eight possible remainders in modulo 8.

5) Starting with 0, you are going to connect each value of x (from the first column in Table 4.1) to its corresponding $6x\,(mod\,8)$ value (from the third column) using a straight line segment. As an illustration and example, we are showing you the connection for $x = 2$, which we determined to be 4 (see Figure 4.3).

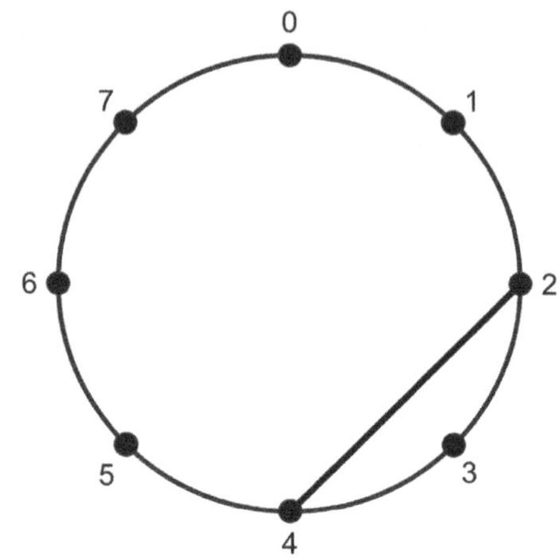

Figure 4.3 A line segment showing the mapping (connection) from 2 to 4 that shows that $6 \times 2 \pmod 8 \equiv 4$ from Table 4.1.

Notice that just because 2 maps to 4, it does not mean that 4 maps back to 2 even though there is a line connecting them. In fact, if you look back at the table you filled in, you should see that the 4 in the first column maps to 0 in the third column. So, when you reach 4 you will draw a line connecting it to 0. In the case that a number maps back to itself (for example, when $x = 0$), you will not introduce any other connections from that number.

As you continue for the other values of x, describe the special features of your diagram, making sense of the connections you see. You have just created a **modulo circle.**

6) What would happen if we extended the table for more values of x (for example, $x = 8$, $x = 9$, $x = 10$)? How can you explain the result?

Repeat for $k = 3$ by first creating a table similar to Table 4.2. One row has been filled out as an example.

Table 4.2 A table for $3x\,(mod\,8)$.

x	$3x$	$3x\,(mod\,8)$
0		
1		
2		
3		
4		
5	**15**	**7**
6		
7		

Create a modulo circle for $3x(\mathit{mod}\,8)$ following the same method that you used for $6x(\mathit{mod}\,8)$.

7) Notice that 0 and 4 do not connect to any other numbers on your circle. Why do you think this happened?
8) Note other interesting details of your $3x(\mathit{mod}\,8)$ modulo circle, providing possible explanations for each feature you note.

Create a modulo circle in $\mathit{mod}\,8$ for $k = 4$ (in the same way you did for $k = 6$ and $k = 3$). If a table helps you, create a new table first.

9) Notice that each value of x maps onto either 0 or 4. Why do you think this happened?
10) Note other interesting features of your $4x(\mathit{mod}\,8)$ modulo circle, providing possible explanations for each feature you note.

Create modulo circles in $\mathit{mod}\,8$ for $k = 5$ and $k = 7$ in the same way as before.

11) For each k value, pose a question specifically for that modulo circle.
12) Note other unexpected features of each modulo circle.

Compare your observations with a friend.

IV. What else?

Now that you have some experience with specific values of k and x, let's analyze these modulo circles further.

1) Consider $kx(\mathit{mod}\,8)$:
   a) What do all of these modulo circles have in common?
   b) Without filling out a table, predict what the modulo circle for each k below would look like. Sketch the appropriate modulo circle to confirm (or adjust) your prediction.
      i) $k = 0$
      ii) $k = 1$
      iii) $k = 8$
2) Consider $kx(\mathit{mod}\,5)$. Create modulo circles for different values of k.
3) Given the $kx(\mathit{mod}\,n)$ values in Table 4.3
   a) Determine the modulus n and the constant k. What interesting features do you notice? Provide a possible explanation for each feature you note.
   b) Are there different values of k that work? Explain your reasoning.

Table 4.3 A table for $kx \pmod n$ from Question 3.

x	$kx \pmod n$
0	0
1	1
2	2
3	3
4	0
5	1
6	2
7	3

4) Look at the modulo circle in Figure 4.4. Recall from our first modulo circle that the connections are not necessarily bidirectional. A mapping in one direction does not necessarily indicate a mapping in the other. For example, the line joining 3 and 1 could represent a mapping from 1 to 3 or from 3 to 1, or both.

   a) What are possible combinations of values for *k* and *n*?
   b) Which features helped you determine *k* and *n*?
   c) Why are there two numbers that are not connected to any other number?
   d) What other interesting features do you notice?
   e) Are there different values of *k* that work? Explain your reasoning.

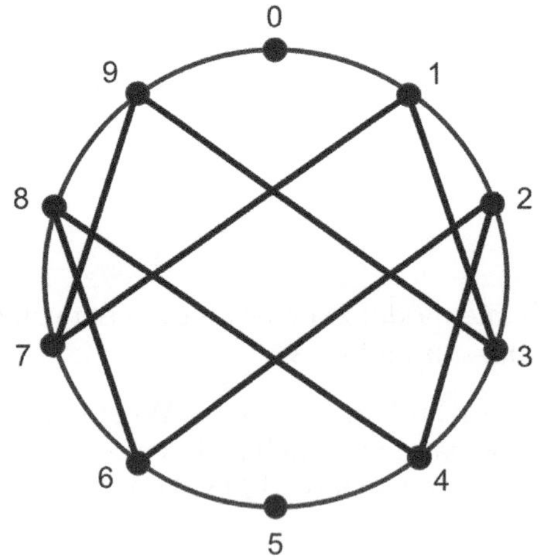

Figure 4.4 A modulo circle for $kx \pmod n$ to be used to determine *k* and *n* in Question 4.

V. Story

Modular arithmetic is relevant to many areas of mathematics including number theory, abstract algebra, and applied mathematics. In this quest, you used the 'wrapping around' feature to show relationships between numbers and create artistic designs. The idea of 'wrapping around' allows us to express relationships (and congruences) of numbers in a different way and opens up a world of possibilities.

In this quest, you explored cyclic relationships of remainders in moduli 8, 12, and 24 using a clock. We are confident that you have encountered other cyclic patterns and modular arithmetic in the world around you. Aside from its mathematical complexity and visually appealing features, modular arithmetic has many useful applications. In fact, ancient scholars used modular arithmetic to analyze planetary movements. In our day-to-day lives, the 'wrapping around' behavior of our seven-day weeks and our 365-day years is impossible to ignore.

Modular arithmetic is central to the mathematics of cryptography, the study of codes. Modular arithmetic is also used in algorithms that help verify identification numbers such as bank routing numbers, credit card numbers, flight ticket numbers, Universal Product Codes (UPC), and International Standard Book Numbers (ISBN).

VI. Going beyond

1) Research the history of modular arithmetic, in particular the contributions from ancient Chinese and Indian mathematics. Create a slideshow of your findings.
2) Create new modulo circles of the form $kx \pmod n$ for different combinations of k and n. Make a visual display of your most beautiful findings.
3) Make an analog clock for someone living on Neptune, where one day is 16 hours long.
4) There are 225 'days' in a year on Venus. Make a calendar for someone living on Venus. Be sure prepared to 'sell' your calendar to others – maybe even to customers from Venus?
5) The Gregorian calendar is used in most parts of the world, but there are many other calendars in use today, like the Chinese, Jewish, or Persian calendars. Explore the patterns in a calendar that you are unfamiliar with.
6) Research how modular arithmetic is used to verify some of the identification numbers mentioned in the 'Story' section.
7) Do research on other applications of modular arithmetic. In what other situations are we concerned mostly with the remainder?
8) Make a video showcasing the mathematics of modulo circles. Set it to music to bring out the beauty and structure of the resulting patterns.

VII. Let's reflect

1) How did this quest challenge you? What new ways of thinking did it encourage?
2) Describe the role of patterns in this quest. How did patterns help you determine how to construct your modulo circles?
3) How would these quests have been different if you had grown up in a world without analog clocks?
4) Can you imagine our lives without the possibility of 'wrapping around' or cyclic behaviors? What would it be like to live in a world without cycles?
5) How does modular arithmetic help us simplify complex problems?
6) What surprised you about this quest? Was there an application, diagram, or calculation that you did not expect to see?

Quest 4.2: Exploring recursive and explicit formulas through visual representations

So many patterns, so little time!

We are surrounded by patterns of repeated arrangements of shapes and objects. Notice the patterns you see around you. Patterns can be found in mathematics too! In Quest 4.1, you explored patterns using modulo circles. In this quest, you will explore the mathematics of patterns further by using numbers, objects, formulas, and diagrams. You will look for patterns in number sequences and diagrams, and make predictions based on the relationships you identify.

I. Invitation

When you think of patterns, what comes to mind? What do you imagine? What characteristics define a pattern?

Look at Figure 4.5. It shows the first four **stages** of a sequence: Stages 1, 2, 3, and 4. The stage number is important because it indicates position in a sequence.

1) What patterns do you see?
2) What do these patterns tell you?
3) How does the shading represent change?

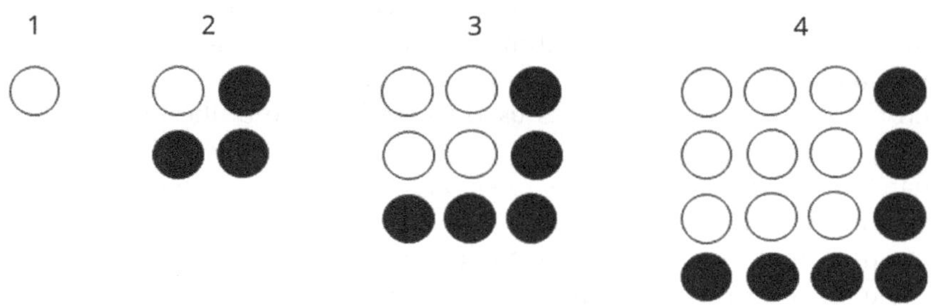

Figure 4.5 The first four stages of a sequence of circles arranged in a pattern.

II. What patterns do you see?

Pascal's triangle is an arrangement of numbers in a triangular shape (see Figure 4.6) that follows a special rule. The 12th row is the last row represented in the figure, but Pascal's triangle continues in this way forever.

1) What patterns do you notice in Pascal's triangle?
2) How do you think Pascal's triangle was created? What is the rule for adding a new row to the triangle?
3) If you were to add a 13th row to our Pascal's triangle, how many numbers would be in it? List the numbers in order.

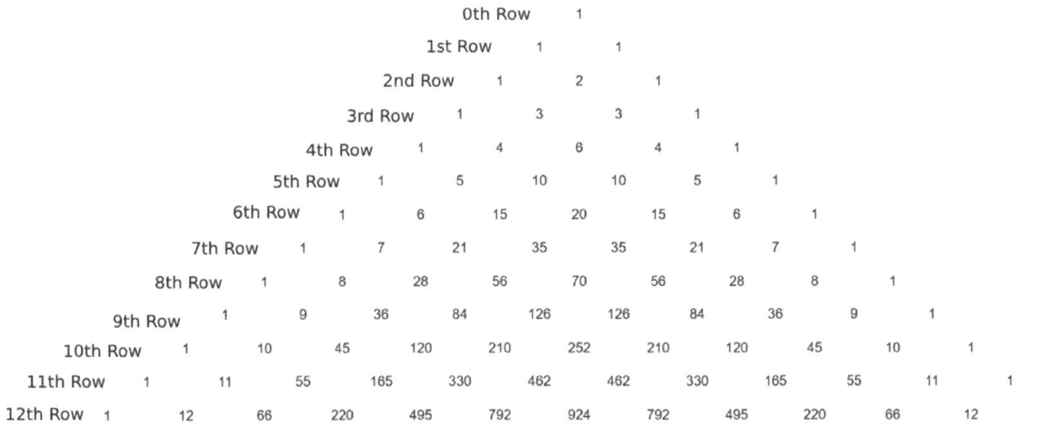

Figure 4.6 Rows 0–12 of Pascal's triangle. Image by Preston Pietruszewski.

In determining how Pascal's triangle was constructed, we can look at how the numbers in a particular row are related to the numbers in the previous row. This is just one relationship to consider.

An ordered list of numbers is called a **sequence**. The numbers in a sequence are called **terms.** When three dots appear at the end of a sequence, it means the sequence continues in the same way and that there are infinitely many terms. Otherwise, the sequence is finite – there is a last term.

Look at the numbers in the third diagonal line of Pascal's triangle (highlighted in Figure 4.7).

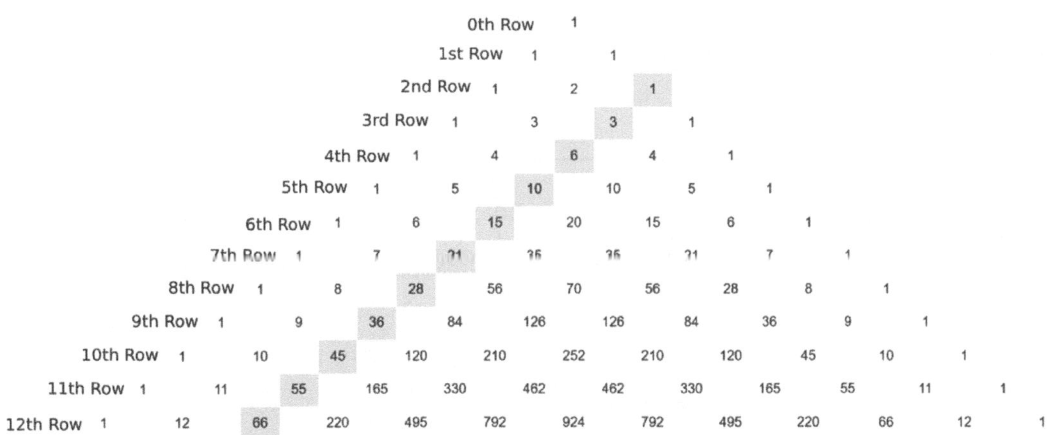

Figure 4.7 Rows 0–12 of Pascal's triangle with the third diagonal highlighted. Image by Preston Pietruszewski.

4) List the sequence of numbers in the diagonal in increasing order, starting with 1.
5) What patterns do you notice in this sequence of numbers?

Find about 60 identical small objects such as coins or beans to represent the first few numbers in the third diagonal: 1, 3, 6, 10, 15, … .

6) How can you arrange your objects to show the relationship between successive terms in the sequence?
7) What new insights does your arrangement reveal about the number sequence?

Jot down any ideas or connections that you notice about the numbers.

The numbers in the infinite sequence 1, 3, 6, 10, 15, . . . are known as **triangular numbers**. Use your objects to determine possible configurations that represent each of the terms in the form of a triangle.

III. Let's see what happens

Patterns can help us model everyday situations. Consider the following problem concerning selfies taken by a group of friends.

Five friends meet at a coffee shop. How many selfies can be taken among these five friends if each selfie includes exactly two friends, and every pair of friends takes exactly one selfie together?

Solve this problem using a method of your choice.

1) How did you solve the problem?
2) How can you represent this problem visually?

Now, let's explore one way we can use patterns to solve this problem. The diagram in Figure 4.8 is a visual representation of the selfies taken by the five friends in the manner described.

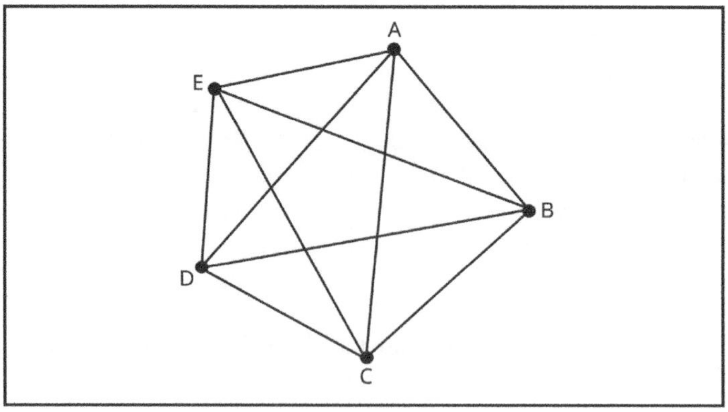

Figure 4.8 A representation of the number of selfies taken by five friends.

3) How does the diagram model the selfies problem? What might the **vertices** (the points labeled with letters) and **edges** (the line segments connecting the vertices) represent?

Let's assume that each labeled **vertex** (single point) represents one of the five friends and that each edge (line segment connecting a pair of vertices) represents a selfie taken between the two friends it 'connects'. (There are other ways of making sense of this diagram, but we will use this interpretation for our example.)

Let's follow this situation chronologically, assuming that each of the five friends (A, B, C, D, and E) arrives at the coffee shop by themself and one at a time.

We want to end up with a diagram like the one in Figure 4.8. We will start with two friends (two vertices). We will call these friends A and B (see Figure 4.9). We are assuming that only A and B have arrived at the cafe so far. This, therefore, is the first selfie to be taken.

Two friends:

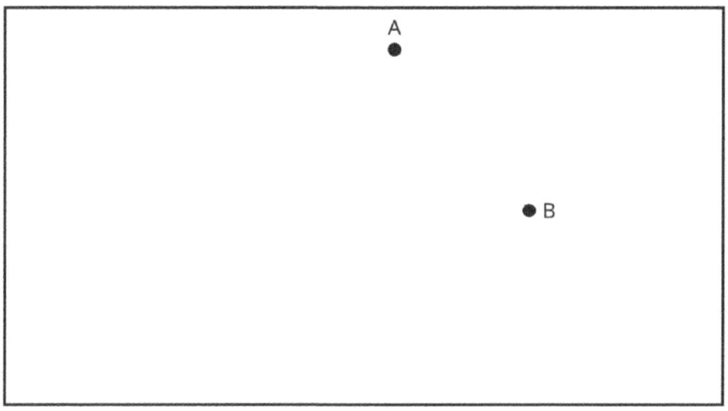

Figure 4.9 Two friends (A and B) represented by vertices.

Connect the vertices with an edge to represent one selfie taken by A and B (see Figure 4.10).

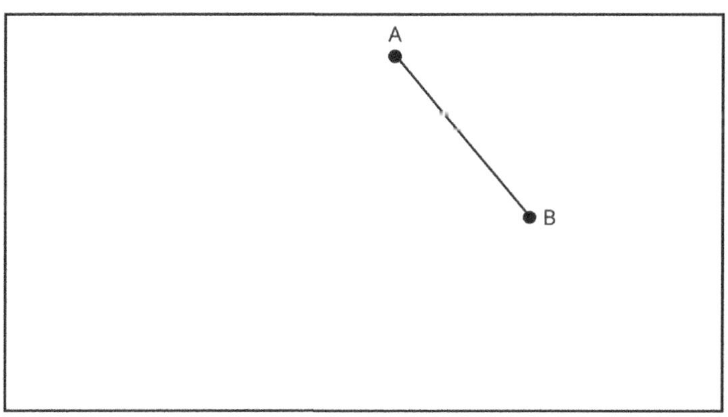

Figure 4.10 A representation of one selfie (an edge) taken by two friends (A and B).

Next, friend C arrives at the coffee shop. Add a new vertex to the existing diagram to represent C (see Figure 4.11). Connect vertex C to A and B.

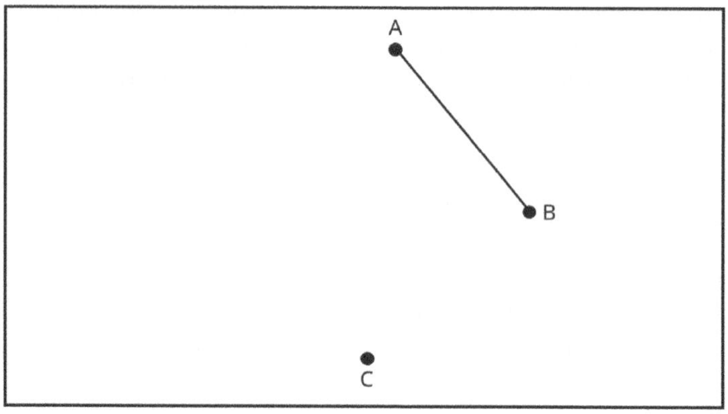

Figure 4.11 A representation of three friends (A, B, and C) as vertices and a selfie between two friends (A and B) as an edge, after C arrives and before any more selfies are taken.

4) How many new selfies were taken after C arrived?
5) How many total selfies have been taken thus far?

Continue to add one vertex at a time and to draw new edges to the existing vertices until you have a diagram with five friends and all of the selfies among them. Count the new selfies as well as the total number of selfies at each stage. Record your findings in a table similar to Table 4.4.

Table 4.4 A table for recording the number of new and total selfies as friends join one at a time.

	2 friends (A and B)	*3 friends (after C joins)*	*4 friends (after D joins)*	*5 friends (after E joins)*
New selfies	1	2		
Total selfies	1	3		

6) What patterns do you notice in the rows showing the number of new selfies and total selfies?
7) How are triangular numbers related to the selfies problem?

Figure 4.12 shows diagrams that represent the number of selfies taken among 2, 3, 4, and 5 friends, where each selfie includes a unique pair of friends. We will call the number of friends x.

8) How are the number of unique two-person selfies taken among x friends and triangular numbers related?

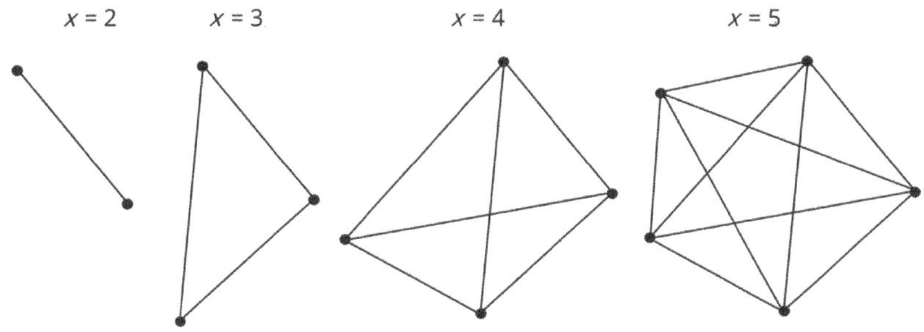

Figure 4.12 Two-person selfies taken among friends where each selfie is taken by a unique pair of friends and where x represents the number of friends. The vertices represent friends and each edge represents a selfie.

IV. What else?

In the selfies case, we were making sense of the problem by considering what was new at each stage. We are going to apply this kind of thinking to explore **recursive formulas**. Recursive formulas help us determine a term in a sequence based on previous terms. This is the kind of reasoning we used in the selfies problem. These formulas are especially useful in situations where understanding *how* things are changing is important because they convey the change occurring at each stage.

Recursive formula for the selfies problem

In our selfies problem, x represented the number of friends. Since two people are needed to take one selfie, our smallest value for x was 2. We can represent the number of selfies taken by two friends as $F(2)=1$. We read the formula as *"F of 2 is equal to 1"*: The number of selfies taken by two friends is one.

We can use a recursive formula to represent the number of selfies taken by any number of friends (under the stated conditions of the problem) if we know that $F(2)=1$. Follow along on the diagrams in Figure 4.12.

- When a third friend joins the group, they add two more selfies to $F(2)$, the number of selfies taken by two friends. We can represent this situation as $F(3)=F(2)+2$. Since we know that $F(2)=1$, $F(3)=1+2$, or 3.
- When a fourth friend joins the group, they add three selfies to $F(3)$, the number of selfies taken by three friends. This is represented by the formula $F(4)=F(3)+3$. We know from before that $F(3)=3$, so $F(4)=3+3$, or 6.

As the number of friends in the group increases, the number of selfies taken with each new person joining the group increases in a particular way. The resulting pattern can help us define the recursive behavior of this problem.

1) In what ways is the selfies problem recursive?

2) Use your observations to write $F(5)$, using $F(4)$. Repeat the process for $F(6)$. What scenario would this represent? Create new diagrams as necessary to help you.
3) As you may have noticed, $F(x)$ depends on the value of $F(x-1)$, the number of selfies taken in the previous stage, before the xth person joins the group. Use this fact to write the recursive formula $F(x)$ for the number of selfies taken by x friends based on $F(x-1)$.
4) Find $F(10)$. How can you make sense of your answer? What scenario would this represent?

Recursive formula for triangular numbers

We will now consider a different representation of a recursive formula for triangular numbers using what we've learned from the selfies problem.

Figure 4.13 shows the first four triangular numbers: 1, 3, 6, and 10. We will use n to represent the stage. In Stage 1 (when $n=1$), we see a representation of the first triangular number, $T(1)=1$. The second triangular number (when $n=2$) is represented by the three circles in Stage 2: $T(2)=3$. As you can see, new circles have been added at each stage to make consecutive triangular numbers. Copy the diagram and color in all circles that represent the new circles at each stage.

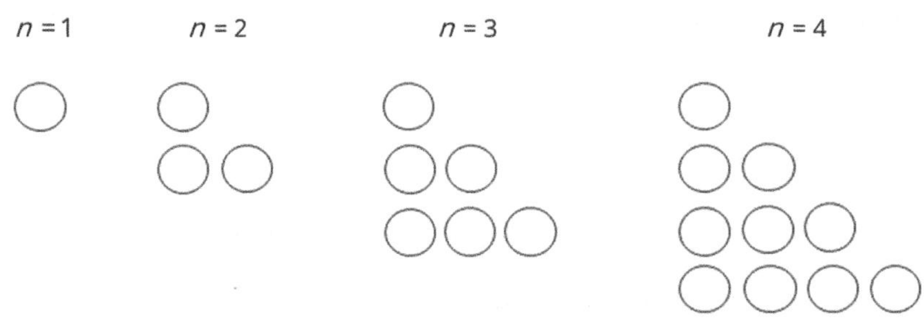

Figure 4.13 Circles arranged to represent the first four triangular numbers (Stages 1–4).

Note that in Figure 4.12, the triangular numbers were represented by the number of line segments (or edges) indicating selfies, but here, the triangular numbers are represented by the number of circles.

5) How many new circles must be added to the fourth triangular number, $T(4)$, to make the fifth triangular number $T(5)$? Draw the arrangement of circles for $T(5)$ and color in the circles you added to $T(4)$ to get $T(5)$. How many circles are in $T(5)$?
6) What is the connection between the stage number (the position of a triangular number in the sequence), n, and the number of circles that are new at that stage?

Like the selfies problem, $T(n)$, the nth triangular number, can be represented using the value of its immediate predecessor, $T(n-1)$. Specifically, $T(1)=1$, $T(2)=T(1)+2$ and $T(3)=T(2)+3$. Note that the first term, $T(1)=1$ in this case, must be known or defined for the recursive formula to work.

7) Write a recursive formula to represent $T(5)$ using $T(4)$. Repeat the process for $T(6)$. Use diagrams as needed.
8) Write a recursive formula, $T(n)$, for the nth triangular number in terms of its immediate predecessor, $T(n-1)$.
9) Find $T(10)$. How can you make sense of your answer?

Notice that the value of $T(1)$ matches the value of $F(2)$ from earlier. But also notice that in $T(n)$, the smallest value of n is 1, and in $F(x)$ the smallest value of x is 2. That is, the first triangular number, $T(1)$, is equal to the number of selfies between *two* friends, $F(2)$.

10) Explain in words why $T(1) = F(2)$.

Recursive formula for square numbers

We will next consider a recursive formula for **square numbers**, the sequence that represents squares of integers: $1^2, 2^2, 3^2, 4^2, \ldots$. Figure 4.14 is a visual representation of the first four square numbers.

Like triangular numbers, square numbers also increase in a particular way at each subsequent stage, n. Copy Figure 4.14 and color in the circles that have been added at each successive stage. This will look familiar to you. We will use $S(n)$ to represent square numbers.

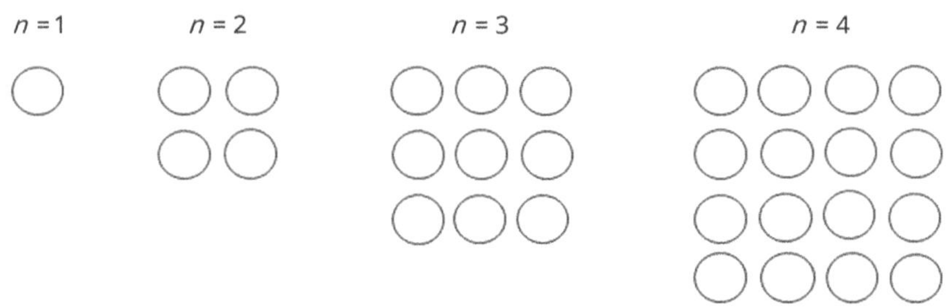

Figure 4.14 Circles arranged to represent the first four square numbers (Stages 1–4).

11) The fourth square number $S(4)$ is 16. How many new circles need to be added to the diagram for $S(4)$ to get the total number of circles for the fifth square number, $S(5)$? Draw $S(5)$ and color in the circles you added to $S(4)$ to get $S(5)$.

This pattern of square numbers can be represented as the following:
$S(1) = 1$, $S(2) = S(1) + 3$, $S(3) = S(2) + 5$ and so on.

12) Describe this pattern using words.
13) Write a formula to represent $S(5)$ using $S(4)$. Repeat the process for $S(6)$.
14) Write a recursive formula, $S(n)$, for the nth square number in terms of its predecessor, $S(n-1)$. If you would like some hints, follow along:
    a) What kinds of numbers are you adding at each stage?
    b) How are these numbers related to the stage number? Can you find a way to represent them using n?
15) Find $S(10)$. How can you make sense of your answer?

Explicit formula for triangular numbers

In contrast to recursive formulas, **explicit formulas** determine a value based only on *n*, the term's position (what we've been calling stage) in the sequence. Unlike recursive formulas, explicit formulas do not rely on the preceding term(s) of the sequence. Rather, explicit formulas use n as the only input value to generate the output. In our example for square numbers, we defined them at the outset as the numbers $1^2, 2^2, 3^2, 4^2, \ldots$. We could rework this as an explicit formula: $S(n) = n^2$, for n = 1, 2, 3, 4, … .

Let's see if we can find an explicit formula for the *n*th triangular number.

The first four triangular numbers are shown as shaded circles in Figure 4.15. Note that we are using shading differently from the recursive formula investigation that goes with Figure 4.13. In Figure 4.15, the number of shaded circles in the *n*th diagram represents the value of the *n*th triangular number. In the figure, you can see that the first triangular number is 1, the second triangular number is 3, the third is 6, and so on. The shaded circles (indicating the triangular numbers) are part of a rectangle. For example, the second triangular number, 3, is represented as a part of a rectangle that has circles arranged in two rows and three columns (2×3) in Stage 2.

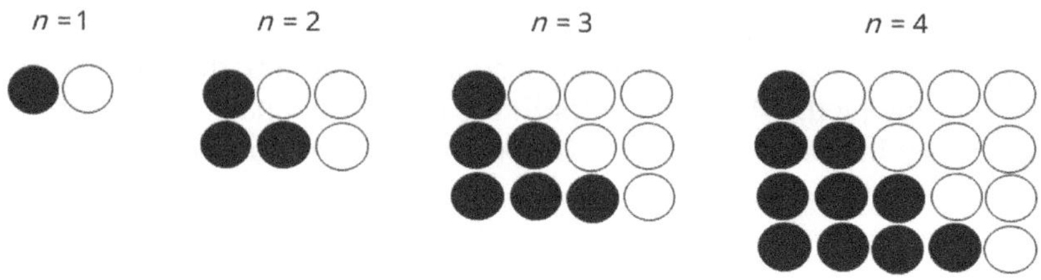

Figure 4.15 Circles arranged in rectangles, where the triangular numbers are represented by the shaded circles.

16) Copy and complete Table 4.5 for $n = 3$ and $n = 4$.
17) What is the connection between the number of circles in the *n*th rectangle and the *n*th triangular number?
18) Draw an arrangement of circles similar to those in Figure 4.15 and shade in the appropriate circles to determine $T(5)$. Repeat the process for $T(6)$.
19) Use your work from Questions 16–18 to help you write an explicit formula, $T(n)$, for the *n*th triangular number.
20) Find $T(10)$. How can you make sense of your answer?

Table 4.5 Table for recording the relationship between rectangular arrangements of circles and triangular numbers.

n	1	2	3	4
The number of rows and columns in the rectangle $r \times c$	1×2	2×3		
The number of circles in the rectangle	2	6		
Triangular number $T(n)$	1	3		

Explicit formula for square numbers

We noted an explicit formula for square numbers, $S(n) = n^2$, when we introduced them earlier in the quest as squares of integers. That is, the *n*th square number is the square of its position, *n*, in the sequence.

$S(1) = 1^2 = 1$
$S(2) = 2^2 = 4$
$S(3) = 3^2 = 9$
$S(4) = 4^2 = 16$
. . .

You are probably familiar with this explicit formula, but let's revisit Figure 4.14 to see it through a geometric representation.

21) How is the explicit formula $S(n) = n^2$ represented in Figure 4.14?
22) What is $S(10)$? How can you make sense of your answer? What would its geometric representation look like?

Another way of thinking about square numbers

Next, you are going to use triangular numbers to derive a different formula for square numbers. Figure 4.16 is a visual representation of this special connection. The corresponding triangular numbers have been shaded.

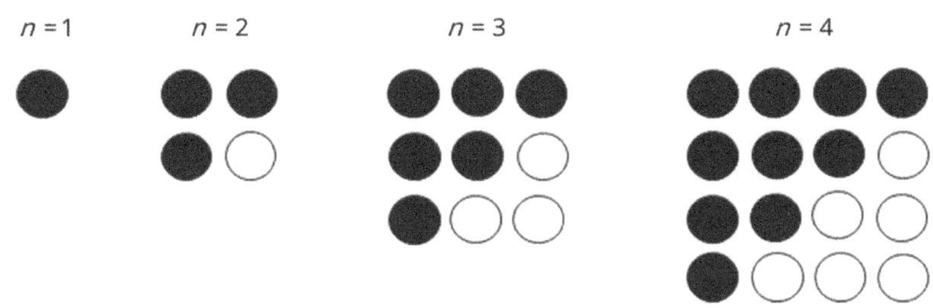

Figure 4.16 Circles arranged in squares, where the triangular numbers are represented by the shaded circles.

23) Copy and extend the diagram in Figure 4.16 to represent the fifth square number, $S(5)$. Color in the circles to show the triangular numbers that make up the square. Repeat this process for $S(6)$.
24) What relationships do you see between triangular numbers and square numbers?
25) Copy and complete Table 4.6 for $n = 3$ and $n = 4$.

Table 4.6 A table for recording the relationship between square numbers and triangular numbers.

n	1	2	3	4
Square number $S(n)$	1	4		
Triangular number $T(n)$	1	3		
Formula for square number, $S(n)$, represented as a sum of triangular numbers	$T(1)$	$T(1)+T(2)$		

26) What is the relationship between the nth square number and nth triangular number?
27) Write a formula for $S(n)$ as a sum of triangular numbers, with the nth triangular number denoted by $T(n)$.
28) Write a formula for $S(n)$ as a sum of triangular numbers, with each triangular number expressed in terms of n. Resist the urge to simplify.
29) Would you consider each of your formulas for $S(n)$ in Question 27 and Question 28 to be explicit or recursive? Why? Can you see someone making an argument for the other case?
30) Find $S(10)$. How can you make sense of your answer?

Explicit formula for the selfies problem

You derived a recursive formula, $F(x)$, for the selfies problem earlier in this quest. Now, let's try to derive an explicit formula for this problem. Recall the diagram representing the number of selfies taken among x friends (see Figure 4.17). Each vertex represents one friend and each edge represents one selfie taken between the two friends it connects.

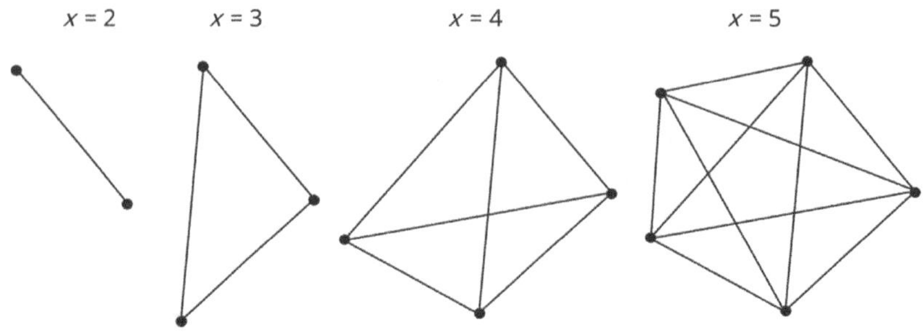

Figure 4.17 Two-person selfies taken among friends where each selfie is taken by a unique pair of friends and where x represents the number of friends. The vertices represent friends and each edge represents a selfie.

31) Draw a diagram to represent $F(6)$, the number of selfies taken among six friends.
32) Draw a diagram to represent $F(7)$, the number of selfies taken among seven friends.

Let's connect what we know about the number of selfies taken among x friends and the formula we developed for the nth triangular number, $T(n)$. Refer to Table 4.7.

Table 4.7 The number of selfies taken among 2, 3, 4, and 5 friends.

Number of friends x	*2*	*3*	*4*	*5*
Number of selfies $F(x)$	1	3	6	10

33) What is the connection between the nth triangular number, $T(n)$, and the number of selfies taken among x friends, $F(x)$?
34) Write an explicit formula, $F(x)$, for the number of selfies taken among x people. It may help to revisit how we determined this for $T(n)$.

Copy and use Table 4.8 to organize some of the formulas you have explored in this quest.

Table 4.8 A table for recording explicit and recursive formulas of triangular and square numbers.

	Explicit formula	*Recursive formula*
The nth triangular number $T(n)$		
The nth square number $S(n)$		
The number of selfies for x friends (if exactly two friends are in each selfie, with no repeated pairings) $F(x)$		

V. Story

Pascal's triangle is an arrangement of numbers wherein each number is the sum of the two numbers diagonally above it. It's a pretty simple concept, right? You will see that this 'simple' arrangement of numbers formed by this straightforward rule contains many surprising relationships and patterns.

In Quest 4.1, you learned about modular arithmetic and about the congruences of numbers in a given modulus. You created visual representations of patterns that result from arithmetic operations on moduli. Modular arithmetic can also be used to create interesting visual patterns using Pascal's triangle. For example, in Figure 4.18, three types of shading (white, gray, and black) are used to represent numbers congruent modulo 3. Note that all circles shaded in black are congruent to 0 $(mod\,3)$; all gray circles are congruent to 2 $(mod\,3)$ and all white circles are congruent to 1 $(mod\,3)$. How would you describe the visual pattern? What would this look like if we were to keep extending Pascal's triangle with more rows? Pascal's triangle is also connected to probability and combinatorics. One connection is to the outcomes of flipping a single coin once, twice, three times, four times, and so on.

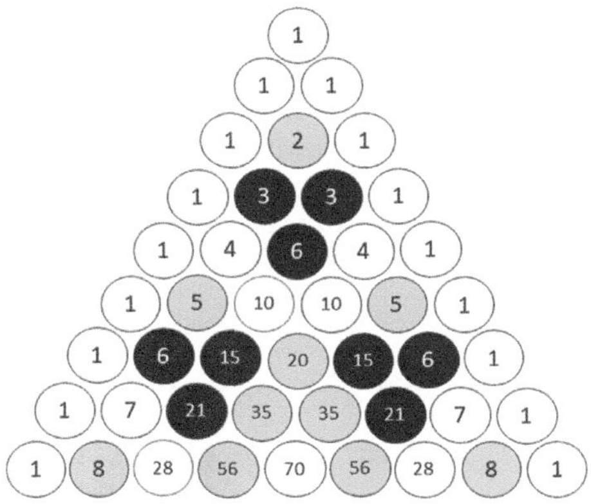

Figure 4.18 Pascal's triangle shaded to show congruence of positive integers modulo 3.

Each time we flip a coin, the possible outcomes are one head or one tail. With a single flip of a coin, there is **one** way to get one head and **one** way to get one tail. Do you see the row that corresponds to the outcome 1–1 in Pascal's triangle?

With two flips, there is **one** way to get two heads (HH), **two** ways to get one head and one tail (HT or TH) and **one** way to get two tails (TT): 1–2–1. There are no other possible outcomes. Do you see the row that has the pattern 1–2–1?

With three coins, there is **one** way to get three heads (HHH), **three** ways to get two heads and one tail (HHT, HTH, or THH), **three** ways to get one head and two tails (HTT, THT, or TTH), and **one** way to get three tails (TTT). Do you see a row with the pattern 1-3-3-1? We summarize the outcomes of flipping coins and their corresponding rows in Pascal's triangle in Table 4.9.

We can use this information to determine various probabilities, such as the probability of flipping three heads in a row. Notice that the flipping of three coins has eight possible outcomes (see the middle column of the third row in Table 4.9). There is only one way to flip three heads in a row. We divide the number of ways to get three heads in a row (1) by the total number of possible outcomes (8) to find that the probability of flipping three heads in a row is ⅛ or 0.125.

Table 4.9 Table showing the possible outcomes of flipping a coin 1, 2, 3, or 4 times with the corresponding rows in Pascal's triangle.

Flips	*Possible Outcomes*	*Integers in successive rows of Pascal's triangle*
1	H T	1, 1
2	HH HT, TH TT	1, 2, 1
3	HHH HHT, HTH, THH HTT, THT, TTH TTT	1, 3, 3, 1
4	HHHH HHHT, HHTH, HTHH, THHH HHTT, HTHT, HTTH, THHT, THTH, TTHH HTTT, THTT, TTHT, TTTH TTTT	1, 4, 6, 4, 1

VI. Going beyond

1) Figure 4.19 shows a pattern of shapes that connects algebra and geometry. Draw the next image in this pattern. How many total shapes (squares and triangles) are there in the 10th image? Suggest an algebraic expression to denote the number of shapes in the nth image.

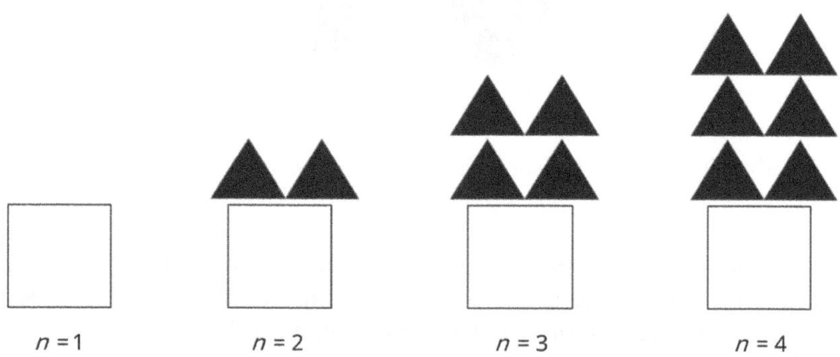

Figure 4.19 Two geometric shapes (square and triangle) arranged to represent an algebraic pattern.

2) If 47 people attend a party and everyone shakes hands with everyone else just once, how many handshakes occur? How is this similar to the selfies problem?
3) The two different shadings in Figure 4.20 indicate numbers that are congruent modulo 2. Create a similar figure, this time using $mod\, 5$. For example, color all numbers congruent to $1\,(mod\, 5)$ using one color and all numbers congruent to $2\,(mod\, 5)$ using another. Repeat for all possible remainders. How is your creation different from the one shown in Figure 4.20 for modulo 2? Do the same exercise for another modulus of your choosing.

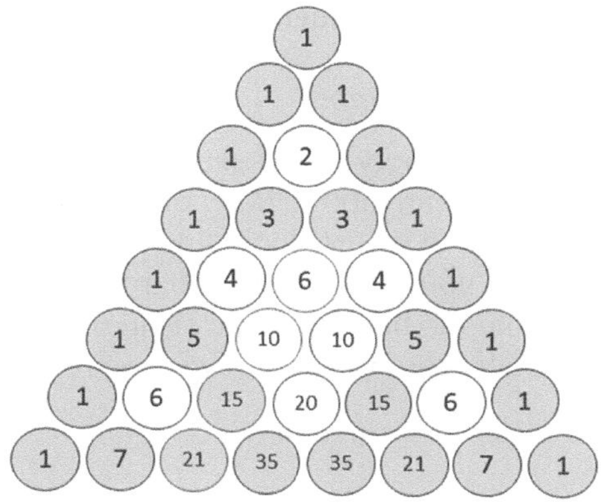

Figure 4.20 Pascal's triangle shaded to show congruence of integers modulo 2.

4) Make and cut out six copies of Pascal's triangles that were colored according to the same modulus you chose for Question 3. Create a kaleidoscope by taping them together such that they meet at the top vertex and form a hexagon. You can even extend the kaleidoscope outward. Take a photo of your artwork.
5) Write an essay about the history of Pascal's triangle.
6) Create your own version of Pascal's triangle. That is, change the rule for making new rows and look for patterns that emerge.
7) Make a 30-second video about patterns involving triangular numbers.
8) Fibonacci numbers are represented by this sequence: 1, 1, 2, 3, 5, 8, 13, … .
   a) What is the next term? What do you notice about the sequence? Look for a sequence of Fibonacci numbers in Pascal's triangle.
   b) Divide each term in the Fibonacci sequence by the previous term. For example, divide 1 by 1, then divide 2 by 1, 3 by 2, 5 by 3 and so on. Keep doing this until you notice something interesting. What do you notice?

VII. Let's reflect

1) What mathematical patterns in this quest did you find to be the most interesting? Most beautiful?
2) How are square and triangular numbers similar? How are they different?
3) In this quest, you identified explicit and recursive formulas. What are some advantages of each type of formula? What are some disadvantages?
4) What questions remain for you after exploring this quest? In other words, what are you now wondering about?
5) What careers depend heavily on the ability to notice and identify patterns? Describe how patterns are used in those jobs.
6) In this quest, you were prompted to see certain mathematics topics that you were already familiar with from a different perspective. For example, whereas you previously understood square numbers to be represented as x^2, you came to understand them also as sums of triangular numbers. Recall an instance, whether in mathematics or another subject, when you were challenged to see a familiar topic from a different perspective.
7) You explored triangular numbers in several contexts: in Pascal's triangle, for determining the number of unique selfies, and as geometric configurations of circles. In what ways do these connections illuminate the nature of mathematics?
8) What part of this quest called on you to be creative?

Quest 4.3: Appreciating self-similarity through infinity

What do broccoli and trees have in common?

In this quest, you will explore the beauty of self-similar shapes known as fractals. You will examine their intricate shapes and learn about their connections to patterns. There are many fractal-like characteristics reflected in nature, including ferns, river tributaries, and lightning paths. We hope this quest helps you notice self-similarity and fractal-like behavior in the world around you and appreciate the beauty of fractals in unexpected places. After becoming acquainted with some fractals, you will design your own unique self-similar object that is beautiful to you.

I. Invitation

Typically in geometry, we learn about straight lines, flat planes, and solid shapes. We learn to measure objects using rulers, protractors, and other measurement tools. We describe length, width, height, and angles of objects and compare them to each other. Some objects are not easy to measure with these tools. Think of an apple. How would you measure its surface area? Objects can also have jagged edges and porous surfaces that can be difficult to measure. Look closely at the crown of broccoli and the tree in Figure 4.21.

1) What features of a tree can you measure easily?
2) What features of a tree are harder to measure than others?
3) What do trees and broccoli have in common?

Figure 4.21 A crown of broccoli (a) and a tree (b) with self-similar characteristics.

We can compare two or more objects based on their shape and size. Two identical objects, with the exact same size and shape, are **congruent**. Two objects are congruent when they overlap perfectly when placed on top of each other, like two 13 cm by 18 cm picture frames.

When two or more objects are **similar,** they share the same shape but not necessarily the same size. For example, two picture frames, one measuring 13 cm by 18 cm and the other measuring 26 cm by 36 cm, are similar because their respective lengths and widths are proportional – the dimensions of the larger frame are twice those of the smaller frame. Note that two congruent objects are also similar since they are the same shape and their dimensions are proportional. Look at your surroundings and find two similar objects. How do you know that your objects are similar to each other?

When an object is similar to one or more of its own parts, we say that the object has **self-similar** characteristics. This means that a small part of the object resembles the whole object. The small part is proportional to the whole. Many objects in nature have self-similar characteristics. For example, stems of a dicentra plant can resemble the larger plant (see Figure 4.22).

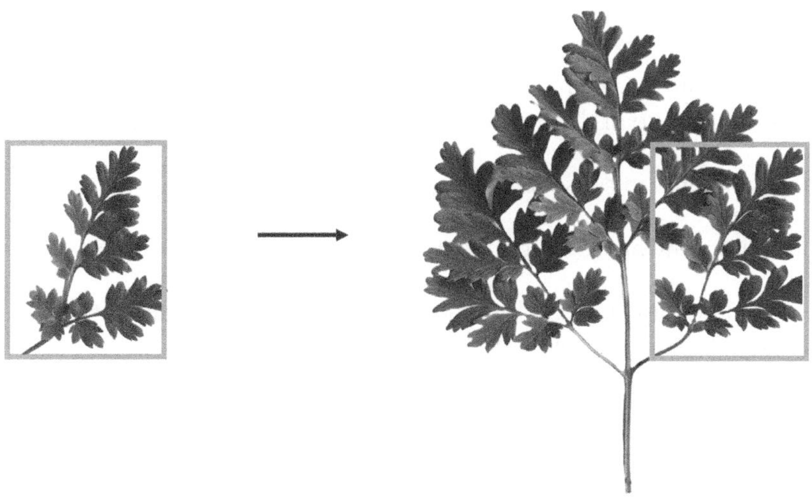

Figure 4.22 A dicentra plant with self-similar parts.

4) In what ways does a tree have self-similar properties? What about broccoli?

II. What do you see in a fractal?

A **fractal** is a shape or pattern whose parts resemble the whole shape. Ferns, broccoli, and trees show these properties in nature. Some fractals can be described through their stages, starting with the initial Stage 0 and continuing through Stages 1, 2, 3, and so on. In this quest, a **code** is a rule that is used to create subsequent stages of a fractal in an iterative manner. We will explore two fractals that use different codes: the Koch curve and the Sierpinski triangle.

Koch curves are fractals that can be constructed by repeatedly modifying a line segment in a particular way. Figure 4.23 shows the first three stages of the Koch curve.

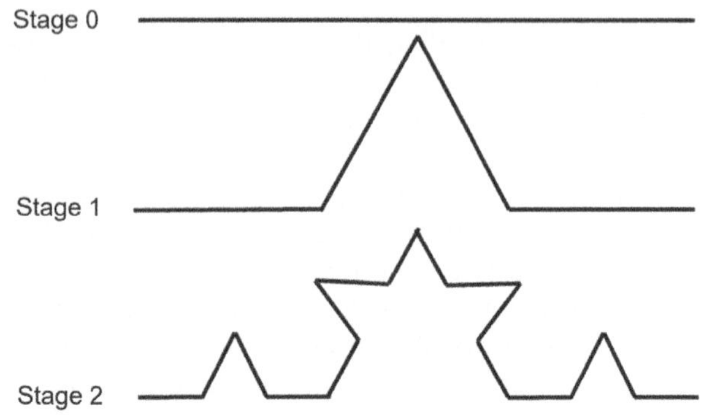

Figure 4.23 Stages 0, 1, and 2 of the Koch curve.

1) Draw Stage 3.
2) What code is being used to construct the fractal? That is, what 'rule' is being applied at each stage to create the next?
3) Describe the structure of the fractal. What is self-similar about it?

Imagine the Koch curve in subsequent stages.

4) Use words and drawings to describe what you imagine.

Imagine what would happen if each of the three sides of an equilateral triangle was a Koch curve (see Figure 4.24 for an equilateral triangle.)

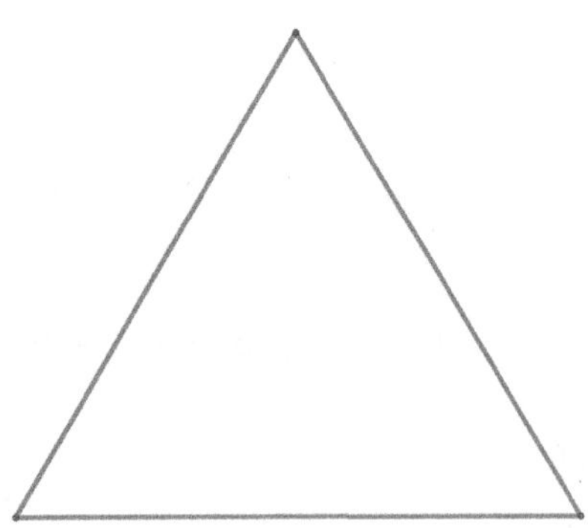

Figure 4.24 An equilateral triangle.

5) Use words and drawings to describe what you imagine. If you had to give the final product a title, what would it be?

The **Sierpinski triangle** consists of shaded and unshaded equilateral triangles. The first two stages of the Sierpinski triangle are shown in Figure 4.25.

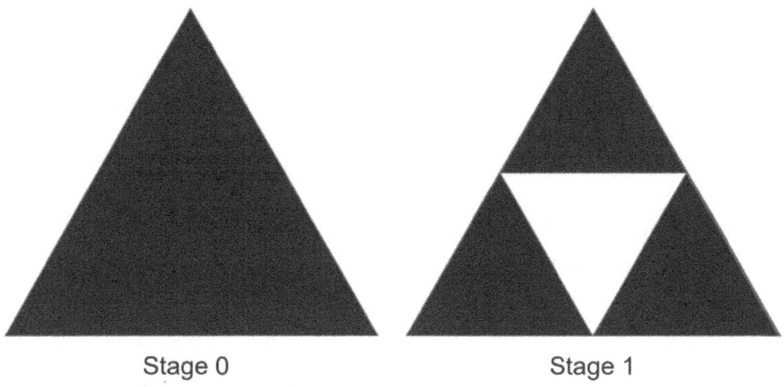

Figure 4.25 Stages 0 and 1 of the Sierpinski triangle.

6) Draw Stages 2 and 3.
7) What code is being used to construct the fractal?
8) Describe the structure of the fractal. What is self-similar about it?

Imagine the Sierpinski triangle in subsequent stages.

9) Use words and drawings to describe what you imagine.

Study the pattern in Figure 4.26 and draw Stage 2 based on the code represented in Stages 0 and 1.

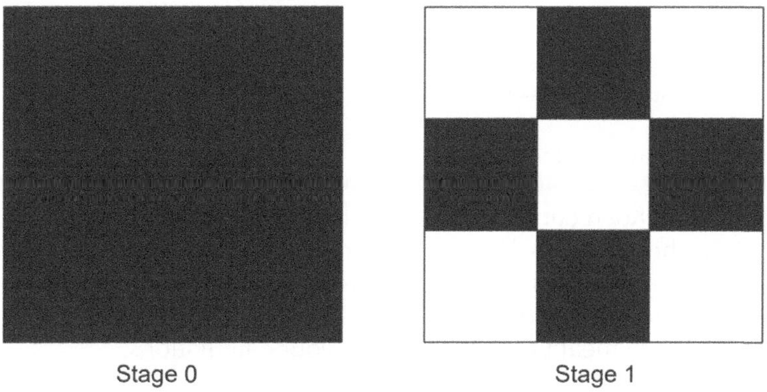

Figure 4.26 Stages 0 and 1 of the new pattern.

Imagine what will emerge in subsequent stages.

10) Use words and drawings to describe what you imagine.

Your next task is to design a self-similar object that is entirely your own. You will use the fractals described in this quest as inspiration for your own unique design.

III. Let's see what happens

Create a fractal using a code of your design. Consider using technology that allows you to scale (make bigger or smaller), copy, rotate, and paste virtual images. If you'd like to take a break from technology (something we all need to do sometimes!), you can use tools like glue, clear tape, colored pencils, markers, and color or poster paper.

1) Create your own Stages 0 and 1. Extend to Stages 2, 3, and 4.
2) Describe what would happen if you could repeat this process over and over. What would happen as the stage numbers get larger and larger, approaching infinity?
3) What makes your fractal beautiful?

Show your fractal to a classmate and ask them to guess your code. Guess their code as well.

IV. What else?

Length of the Koch curve

Consider Stages 0, 1, and 2 of the Koch curve, shown again in Figure 4.27.

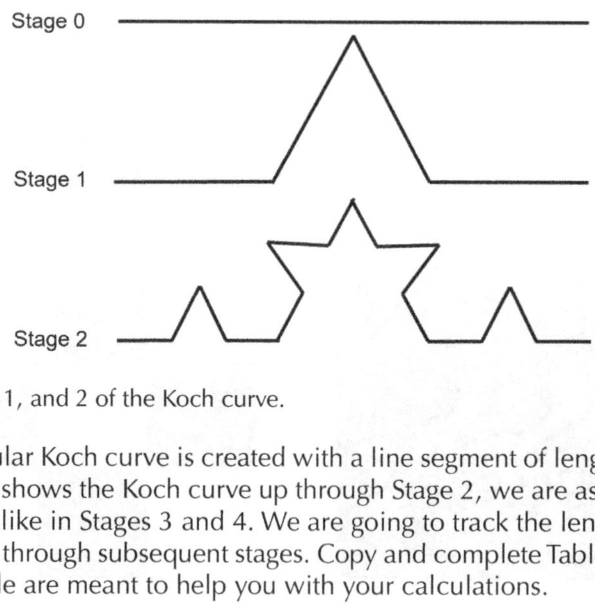

Figure 4.27 Stages 0, 1, and 2 of the Koch curve.

Let's say this particular Koch curve is created with a line segment of length 27 at Stage 0. Although Figure 4.27 shows the Koch curve up through Stage 2, we are asking you to imagine what it would look like in Stages 3 and 4. We are going to track the length of each segment and the total length through subsequent stages. Copy and complete Table 4.10. The numbers provided in the table are meant to help you with your calculations.

Table 4.10 Table for recording data related to successive stages of the Koch curve.

	First five stages of the Koch curve				
	0	*1*	*2*	*3*	*4*
Number of segments	1	4			
Length of each segment	27	9			
Combined length of all segments	27	36			

1) What patterns did you notice in your table and how did they help you complete it?
2) What predictions can you make about the values in the stages beyond Stage 4?
3) What do the values in each row tell you about the fractal?

You will next explore the patterns in the areas and perimeters in the successive stages of the Sierpinski triangle.

Area of the Sierpinski triangle

Consider the equilateral triangle that represents Stage 0 of the Sierpinski triangle (see the left side of Figure 4.28). This triangle has an area of 64 square units. Figure 4.28 shows you Stages 0 and 1 of the Sierpinski triangle. We are asking you to imagine subsequent stages.

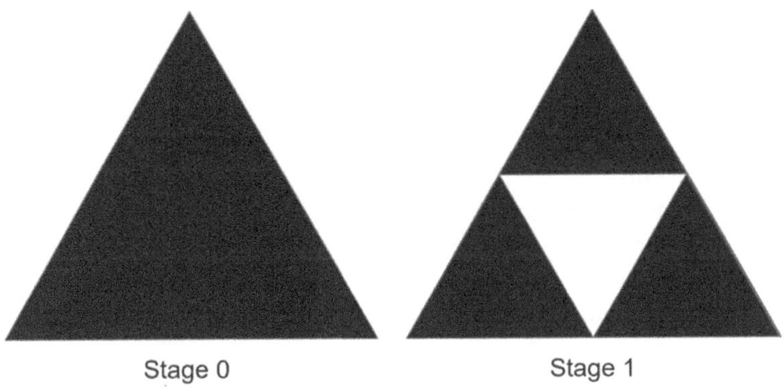

Figure 4.28 An equilateral triangle with an area of 64 square units at Stage 0, on the left. The figure on the right represents Stage 1 of the Sierpinski triangle.

What is the area of the shaded portion of the Sierpinski triangle in successive stages? Copy and complete Table 4.11. As you fill it out, look for patterns in each row and column.

Table 4.11 Table for recording data related to the area of successive stages of the Sierpinski triangle.

	First five stages of the Sierpinski triangle				
	0	*1*	*2*	*3*	*4*
Number of shaded triangles	1	3			
Area of each shaded triangle	64	16			
Total area of shaded triangles	64	48			

4) What patterns did you notice in your table and how did they help you complete it?
5) What predictions can you make about the values in the stages beyond Stage 4?
6) What do the values in each row tell you about the fractal?

Perimeter of the Sierpinski triangle

Next, let's explore patterns in the perimeter of the Sierpinski triangle. At Stage 0, we start with an equilateral triangle whose perimeter is 24 units (see the left side of Figure 4.29).

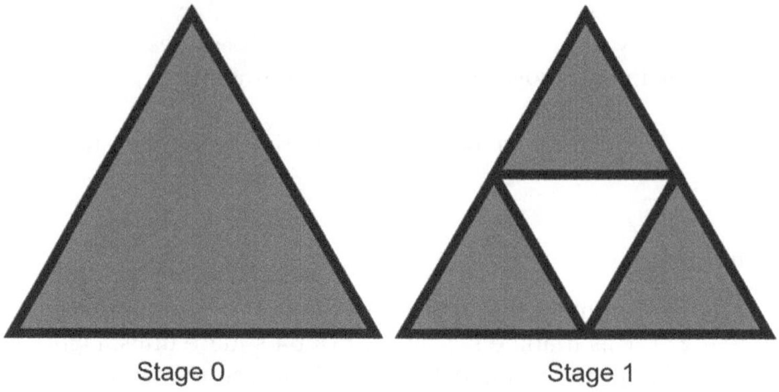

Figure 4.29 An equilateral triangle with a perimeter of 24 units at Stage 0, on the left. The figure on the right represents Stage 1 of the Sierpinski triangle.

Note that this equilateral triangle is different in size from the one in Figure 4.28. Again, we are showing just the first couple of stages and asking you to imagine what follows.

Copy and complete Table 4.12. As you fill it out, look for patterns in each row and column. Use Figure 4.29 to guide your thinking. Although perimeter usually refers to the total length of the outside of a figure, here we will take perimeter to be the combined length of the outside of all of the 'small' triangles. Use Figure 4.29 to help you get started. Note that the total perimeter of all shaded triangles in Stage 1 is 36.

Table 4.12 Table for recording data related to the perimeter of successive stages of the Sierpinski triangle.

	First five stages				
	0	*1*	*2*	*3*	*4*
Number of shaded triangles	1	3			
Perimeter of each shaded triangle	24	12			
Combined perimeter of all shaded triangles	24	36			

7) What patterns did you notice in your table and how did they help you complete it?
8) What predictions can you make about the values in the stages beyond Stage 4?
9) What do the values in each row tell you about the fractal?

Geometric sequences in fractals

A **geometric sequence** is an ordered list of numbers that can be written as

$$a, ar, ar^2, ar^3, ar^4, \ldots$$

Notice that the first term in the sequence is represented by the letter a and each term is being multiplied by the same number, r, to determine the next term. This number, r, is called the **common ratio**. For example, the sequence 1, 2, 4, 8, . . . is a geometric sequence whose first term is 1 and whose common ratio is 2.

Sequences whose terms approach a specific finite value **converge**. When a sequence does not converge, it **diverges**. For example, when the terms of a sequence get bigger and bigger and approach infinity, the sequence diverges. The sequence 1, 3, 9, 27, 81, 243, . . . is an example of a divergent geometric sequence with $a = 1$ and $r = 3$. Notice that the first term is 1 and that each term is being multiplied by 3 to determine the next term. The sequence $1, \frac{1}{2}, \frac{1}{4}, \frac{1}{8}$, . . . is an example of a convergent geometric sequence. The first term, a, is 1 and the common ratio, r, is $\frac{1}{2}$. Notice that the successive terms get closer and closer to 0.

10) What geometric sequences do you see in Table 4.10? In Table 4.11? In Table 4.12?
11) Which of these sequences diverge? How can you tell that they diverge?
12) Which of these sequences converge? How can you tell that they converge?

V. Story

One of the most beautiful fractals in modern mathematics is the Mandelbrot set (see Figure 4.30). This famous fractal is named after Benoit B. Mandelbrot, the mathematician who displayed the set using computer technology in the late 1970s. What used to be difficult to imagine was now on full display, to be admired by everyone. Mandelbrot coined the phrase 'fractal' and popularized the field of fractal geometry. Since the 1970s, fractals have gained immense popularity and have been applied in a variety of fields outside of mathematics, including medicine, biology, engineering, economics, technology, music, computer graphics, fashion, and art.

The Mandelbrot set, like other fractals, has a 'zooming in' feature, with each magnification revealing new details (and surprises!) For example, when zooming into areas near

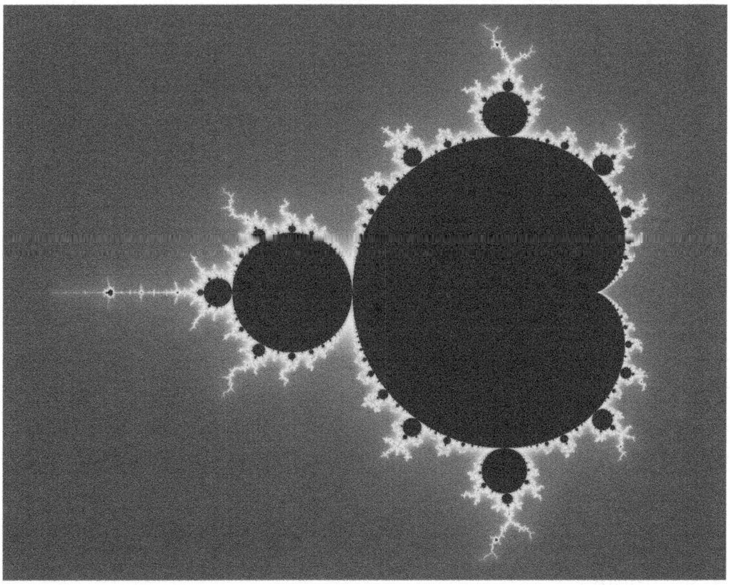

Figure 4.30 The Mandelbrot set. Image by Wolfgang Beyer retrieved from https://commons.wikimedia.org/wiki/File:Mandel_zoom_00_mandelbrot_set.jpg.

the boundary of the set, we come across shapes that look like the set itself – with some variation. We encourage you to find an image of the Mandelbrot set on the Internet. You will notice the different colors surrounding the boundary of the set. Different points on and off the set are colored based on the behavior of each point in the iterative process defined by the mathematics of the set – specifically, on their long-term convergent or divergent behavior. Search for images and information about this set to learn more about its beauty and complexity, as well as the mathematics behind it. Find interactive websites or watch videos that allow you to zoom in to find parts of the set that look like the whole. Look at the colors and notice some other interesting patterns. You are in for a special treat!

VI. Going beyond

1) Take a photograph of a self-similar object and explain what makes the object self-similar.
2) What geometric sequence do you see in Figure 4.31?

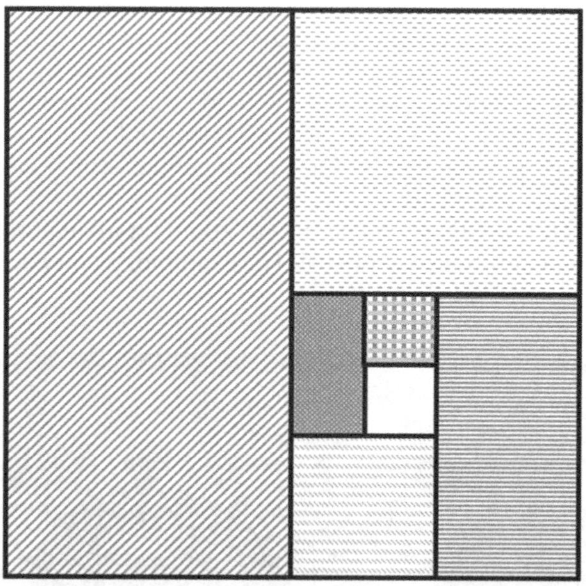

Figure 4.31 A visual representation of a geometric sequence.

3) Modify the code of one of the 'classic' fractals in this quest: the Koch curve or Sierpinski triangle. Show the first four stages of the modified fractal.
4) Make a scrapbook of fractal-like behavior you see in nature using photographs, drawings, or actual artifacts (such as dried plants). Categorize the images based on their shared features.
5) Fractals are used in many fields including art, music, medical research, and technology. Find out more about how fractals are used in your field of study or a related field and summarize your findings.
6) Figure 4.32 shows Stages 0 and 1 of three different self-similar designs, each with its own distinct code. Describe the code and draw Stage 2 for each design.

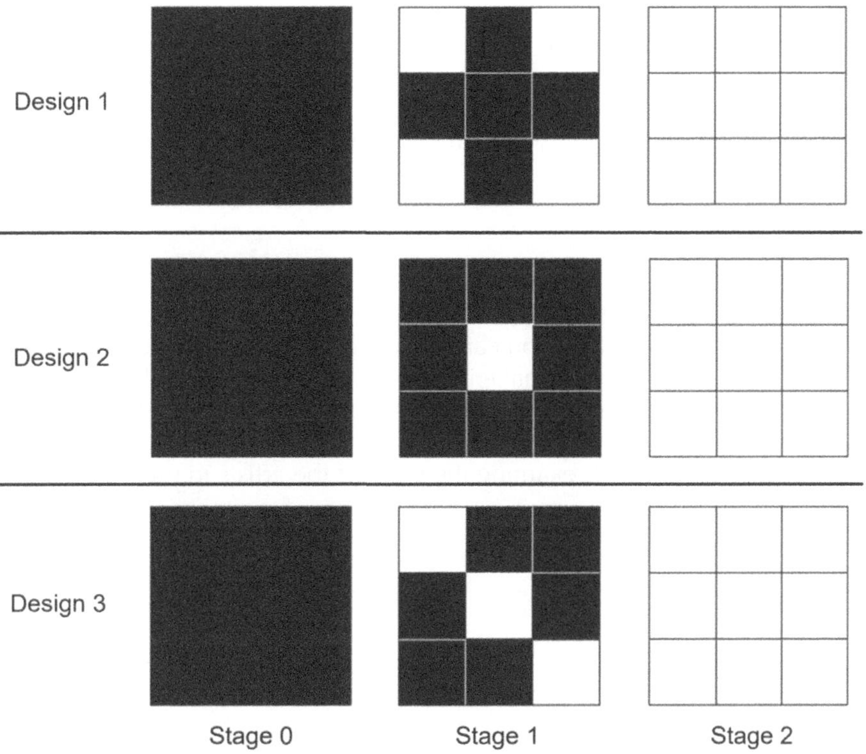

Figure 4.32 Stage 0 and 1 of three designs for Question 6.

VII. Let's reflect

1) How does the concept of fractals challenge your previous notions of geometry?
2) In what ways are iterative processes present in your everyday life?
3) How do fractals help you better understand the world around you?
4) Write a poem that captures the essence of fractals.
5) Are fractals only visual? Can fractals be captured through sound or other senses? How else can you 'experience' a fractal?
6) Have you ever felt like you were in a fractal situation? Describe the situation.

Project 4: Math museum of patterns

This book challenges you to find mathematics in unexpected places and situations, like when you are juggling, doodling, or grocery shopping. Are you ready for yet another challenge of this sort? This project will have you visit a museum and interact with its exhibits using mathematics as a lens. You will then create a museum exhibit that features a beautiful mathematical pattern.

Part I: Museum visit

For this project, you will visit a museum. It may be a museum that you don't initially associate with mathematics: That's fine. You can visit a museum you have been wanting to see for a while. This is your opportunity to make it happen. Check your local library for free museum passes. Make a day of it! (If you cannot visit a museum in person, find a museum online and take a virtual tour.)

Prior to your museum visit, examine the work of the artists in this list, looking for mathematics in their work:

- Yayoi Kusama
- M.C. Escher
- Leonardo da Vinci
- Sol LeWitt
- Gustav Klimt
- Oswaldo Guayasamin
- Lygia Clark
- Thandiwe Muriu

How do these artists use mathematical patterns in their work?

On your visit, interact with the museum's exhibits using mathematics as a lens. Look for mathematical patterns; that is, find evidence of regularity or repetition in the structure, design, or objects at the museum. Can you detect the use of patterns in the artists' techniques? For each mathematical pattern that you find, ask yourself the following questions: What makes it a pattern? What makes the pattern mathematical? In what way is this pattern beautiful?

Find as many mathematical patterns as you can. Look for patterns in places where there appears to be no patterns. Write a one-page reflection paper from your museum visit. In your reflection, respond to the following prompts and include sketches or photos where you can.

1) What museum did you visit?
2) Did the museum connect to your interests or discipline of study? In what way?
3) Describe the patterns you found. Include photos or sketches and refer to them in your descriptions.
4) What questions would you ask the artists of these works if you could meet them?
5) What will you take away from the visit related to mathematics and creativity?
6) What question(s) would you like to pursue further?

Part II: Your exhibit

Throughout this book, and especially in Chapter 4, you have had a chance to consider the beauty of mathematics through patterns. Build on this experience by creating your own design that is just as mathematical as it is creative. Design a mathematical pattern whose formula or rule can be repeated by someone else. Your exhibit can be connected to any field of mathematics (number theory, algebra, arithmetic, geometry, calculus, etc.). Your instructor may ask you to display your exhibit in your classroom or online. Feel free to also share it with your friends or submit it to be considered for a display at a local art center.

The suggestions for what you produce are left intentionally vague. Here are the basic requirements for your exhibit:

1) It must incorporate a mathematical pattern.
2) It should include a placard that has a title, your name, and a short description of the pattern. (Think of museum displays.) The description should be concise and fuel curiosity, so try to refrain from being overly descriptive. Let the audience find the pattern in your exhibit, rather than be told where or how to find it.
3) Your exhibit must be an item for display. Here are some exhibit ideas (but there are many more possibilities that you can consider):
    a) Poster
    b) Video
    c) Art piece such as a painting, sketch, or sculpture
    d) Audio file
    e) Application or computer program you develop
    f) Live performance
    g) Music piece
    h) Interactive demonstration
    i) Book or a booklet
    j) Children's book
4) Your final product should be
    a) Connected to a mathematical pattern
    b) Thorough and accurate (particularly in its use of mathematics)
    c) Innovative – introducing a new mathematical pattern (not a replication of an existing design or diagram)
    d) High quality – polished enough to be displayed in a public forum

Part III: Reflection on your exhibit

Write a one-page summary about your exhibit. In your report, respond to the following prompts:

1) What was your inspiration or motivation for the exhibit?
2) What mathematical pattern did you explore? Explain the mathematics of your project.
3) What did you learn?
4) How does the project connect to your interests or discipline?
5) If given the opportunity how would you change, add to, or improve your exhibit?
6) What question(s) would you like to pursue further?

Chapter introduction: Cited and related works

Cropley, D., & Cropley, A. (2008). Elements of a universal aesthetic of creativity. *Psychology of Aesthetics, Creativity, and the Arts, 2*(3), 155–161. https://doi.org/10.1037/1931-3896.2.3.155
Kaufman, J. C., & Beghetto, R. A. (2009). Beyond big and little: The Four C model of creativity. *Review of General Psychology, 13*(1), 1–12. https://doi.org/10.1037/a0013688

Quest 4.3: Cited and related works

Naylor, M. (1999). Exploring fractals in the classroom. *The Mathematics Teacher, 92*(4), 360–366.
Peitgen, H.-O., Jürgens, H., Saupe, D., Maletsky, E., & Perciante, T. (1999). *Fractals for the classroom: Strategic activities* (Vol. 3). Springer.
Sobel, M. A., & Maletsky, E. M. (1999). *Teaching mathematics: A sourcebook of aids, activities, and strategies* (3rd ed.). Allyn and Bacon.

Project 4: Cited and related works

Munakata, M., Monahan, C., Krupa, E., & Vaidya, A. (2022). Non-traditional assessments to match creative instruction in undergraduate mathematics courses. *International Journal of Mathematical Education in Science and Technology, 54*(7), 1272–1287. https://doi.org/10.1080/0020739X.2022.2128452

5 What makes you wonder?

This chapter closes out your journey with *Creative Thinking Through Mathematics – the book* – but does not close out your journey with creativity through mathematics – *your life*. Think of this chapter as a springboard for further explorations. The quests in this chapter are meant to guide you as you wonder about your world. We want these quests to do more than pose mathematical questions. We want them to make you think about what else you could be wondering about from a mathematical perspective. You will undoubtedly continue to encounter mathematics in your life: We invite you to view each situation or problem through the lens of creativity. Mathematics is a dynamic, ever-growing field of study, and we hope that the open-ended nature of topics in this chapter reassures you that your personal discoveries in mathematics – your personal 'little-c's – matter.

As you've experienced in the previous quests, mathematics can lead to unexpected discoveries. This chapter will be no different. We all learned about triangles in school, but is there more to a triangle? In what ways can maps reflect a privileged ideology? How might maps exclude a group of people? Asking open-ended questions (Acar et al., 2023) and being comfortable with ambiguity (Zenasni et al., 2008) are two key features of creativity. The quests in this chapter will undoubtedly open more doors than they close. That is our intention. We embrace a growth mindset (Dweck, 2006), where learning is ongoing, and view learning as a continual process. Specifically, we hope that this book encourages you to continue to wonder about the world around you through a mathematical lens.

These quests will test your fundamental understandings of the world and of mathematics: How many sides does a piece of paper have? Must the angles in a triangle add up to 180°? Are maps accurate? With these questions, we invite you to challenge conventions and see what else is possible!

Chapter 5: What makes you wonder?

DOI: 10.4324/9781003541097-6

Quest 5.1: Surprising encounters with the Möbius strip

Is there such a thing as one-sided paper?

In this quest, we challenge your notions about properties of geometric objects. When we describe geometric objects such as triangles, cubes, and pyramids, we can talk about their features, including their surfaces, sides, and edges. In this quest, you are going to use strips of paper to reexamine and challenge your previous understanding of geometry. Your ability to question, predict, explore, and defend new ideas will play a key role. You will have opportunities to ask "But how?" and "What if?". We invite you to pursue such questions.

I. Invitation

We begin this quest with a rectangular strip of paper (see Figure 5.1). Despite its simple shape, a rectangular strip of paper has many interesting features like length, width, height (or thickness), area, perimeter, and even volume. We can use many tools, including rulers, to study its properties. But these are not the only features that a paper strip can have; this quest will lead you to some surprises.

We recommend that you cut a sheet of standard A4 printer paper (which is about 30 cm by 21 cm) into strips that are each about 30 cm by 2.5 cm, but any long paper strip will suffice. For our examples, we will use the term **side** to refer to the surface of the paper and **edge** to refer to a border of the sides.

Most of us would agree that a piece of paper has two sides – a front and a back. We can also count the edges of a piece of paper. You can see in Figure 5.1 that this strip of paper has four edges.

1) What else can you count or observe about a paper strip?
2) What are different ways in which you can argue that there are two sides to this strip of paper? How would you convince someone?
3) What distinguishes the two sides?
4) Can you imagine a piece of paper that has two edges? How would you construct this from your strip of paper? Try out some ideas and take pictures of your results.

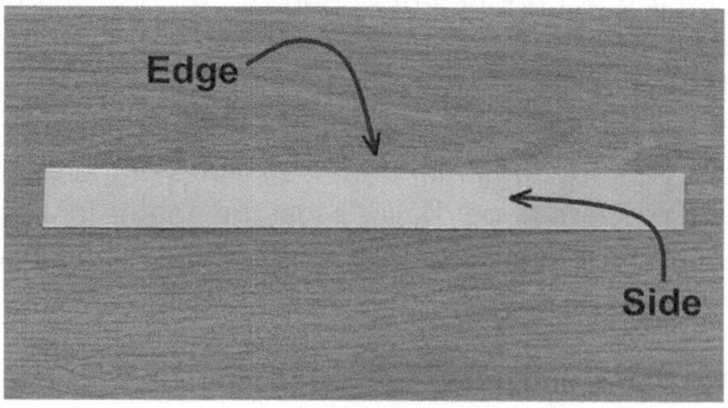

Figure 5.1 A rectangular strip of paper that is about 30 cm by 2.5 cm.

You can manipulate paper strips in interesting ways. They can be cut, folded, twisted, curled or taped together! Answer the following questions about different shapes made with paper strips.

5) Tape the two short ends of one paper strip together to make a loop (see Figure 5.2). You may recall the loops you made in Quest 2.1. What do you wonder about loops now?

Figure 5.2 A loop with the short ends of a strip taped together.

6) How many sides does the loop have? For a strip of paper, we would call the sides 'front' and 'back'. What would you call the sides of the loop you just created?
7) We claim that there are two sides to this loop. How can this claim be justified? Consider coloring each side to demonstrate your thinking.
8) What are some other ways to argue that there are two sides to the loop?
9) The original strip of paper had four edges. You could call these top, bottom, left and right. How many edges does the loop have? What would you call them? Consider coloring each edge to explain your answer.
10) We claim that there are two edges to the loop. Aside from coloring each edge, what are some other ways to justify this claim?
11) What can you say about the length of the loop?
12) Pretend that the strip is a highway. Think about what would be the center divider, which we will call the midline. (Recall a midline just like this in Quest 2.1).
    a) Make a prediction about what would happen if you cut along the midline of the loop you made in Question 5.
    b) Choose a starting point and make a small incision. Cut along the midline of your loop until you return to your starting point (see Figure 5.3). What do you notice about the result? How does the result compare to your prediction? Did anything surprise you about the result?
    c) How many total sides and edges resulted from cutting the loop in this way?

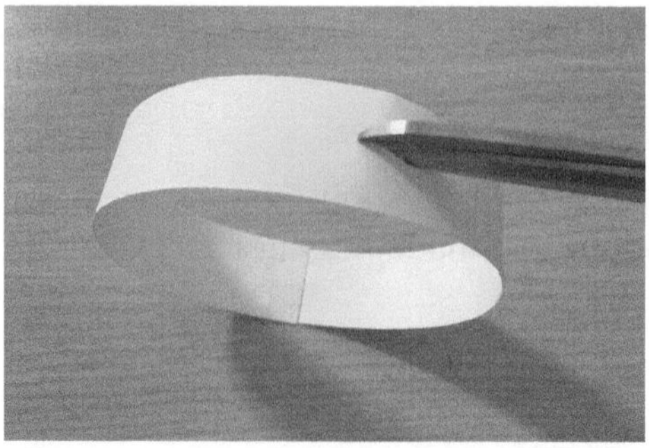

Figure 5.3 Starting your cut along the midline.

II. What makes you wonder?

Take another strip of paper. This time, we want to introduce a half-twist. We recommend making the shape of a loop first (without taping) and then introducing one half-twist at one of the short ends before taping the short ends together (see Figure 5.4).

Figure 5.4 A modified loop with a half-twist (aka Möbius Strip) formed by making a half-twist at one end and then taping the ends together.

1) Now what do you wonder?

Look closely at your object and answer the following questions:

2) What do you notice about the object's shape?
3) How many sides does this object have? How do you know?
4) How many edges does this object have? How do you know?
5) What can you say about the length of the loop after it's been modified in this way? How would you investigate this?
6) How else might you modify this shape to learn something new about it?

Use what you discovered about the number of sides and edges of the modified loop to describe what you created.

III. Let's see what happens

The modified loop you have just created is called a **Möbius strip**. In this quest, you will modify Möbius strips in different ways and explore the results.

Writing on a Möbius strip

Consider a page in a book. If you were reading something written on one page, you would need to turn or 'flip' the page to read what is on the other side. Does this convince you that there are two sides to the page?

Let's see what happens when we write on a Möbius strip.

1) What do you think would happen if we wrote on both sides of a strip of paper before introducing a half-twist and taping the edges together? (Make sure to answer this question before trying it out.)

Go ahead and cut a strip of paper with the same dimensions as before. Write half of your favorite poem, song, or saying on one side of the strip of paper, going across the long way from left to right. Make sure to fill the whole length of that side.

We chose the first four lines from a William Blake poem as an example.

> "To see a world in a grain of sand
> And a heaven in a wild flower,
> Hold infinity in the palm of your hand
> And eternity in an hour."
> – William Blake, 'Auguries of Innocence'
> From *Life of William Blake,* "Pictor Ignotus"
> (Gilchrist, 1863)

We took a paper strip and wrote, "To see a world in a grain of sand and a heaven in a wild flower" on one side of it, filling the whole strip from left to right (see the top photo in Figure 5.5).

Flip the strip along the top long edge and write the second half of your saying on the other side from left to right. It is very important that you flip the paper strip along the length. Be sure to fill the whole length of that side too.

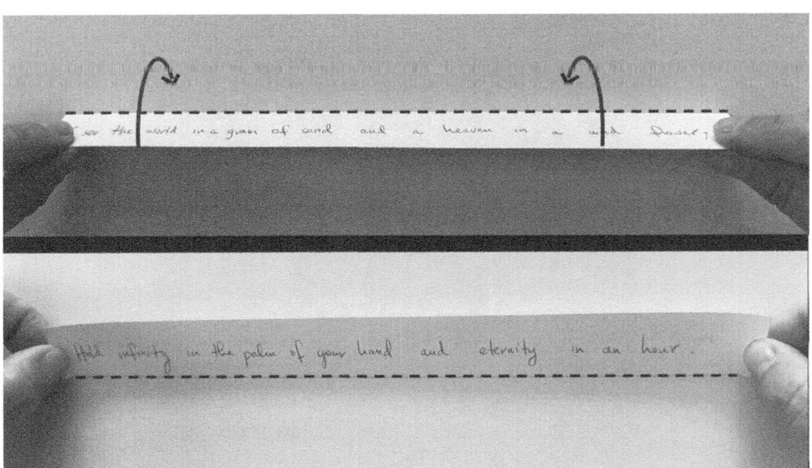

Figure 5.5 Flipping the paper strip along the top long edge: first half of the excerpt visible before the flip (top photograph) and second half of the excerpt visible after the flip (bottom photograph).

For the excerpt we chose, we wrote, "Hold infinity in the palm of your hand and eternity in an hour" on the other side of our strip (see the bottom photo in Figure 5.5.)

Now make a Möbius strip from the strip containing your poem (by introducing a half-twist and taping the short edges together as before) and read your excerpt from the beginning (see Figure 5.6).

Figure 5.6 An image of the excerpt of the Blake poem written on a Möbius strip.

2) What did you observe?
3) How does what you observed help you determine how many sides there are to a Möbius strip?

Cutting a Möbius strip along the midline

Make a new Möbius strip by following the same instructions as before. Pretend that your Möbius strip is a highway, like you did for the loop in Figure 5.3. We are going to cut along the midline. Begin by making a small incision first to mark a starting point (see Figure 5.7).

Before continuing your cut, state your predictions about the resulting object.

Figure 5.7 Starting the cut of the Möbius strip along the midline.

Go ahead and cut along the midline until you return to the starting point.

4) How would you describe what happened?
5) How many sides and edges did you end up with? How did you determine this?

Compare the outcome of cutting a Möbius strip along its midline with what resulted from cutting the original (unmodified) loop along its midline.

6) Why do you think this happened?

A single cut along the midline of the Möbius strip leads to a surprising result. This result is one of many similarly beautiful, and perhaps unexpected, outcomes of explorations of the Möbius strip.

Cutting modifications of the Möbius strip

Two half-twists: Make a new Möbius strip, but instead of introducing one half-twist to the strip of paper as you did for the original Möbius strip, make *two* half-twists (or one full-twist) at one end before taping the ends of the strip together.

Figure 5.8 A modified loop with two half-twists.

7) Does your result look like the image in Figure 5.8?
8) How could perspective or the thickness of the paper you use make your object look different from what is shown in the figure? Is it, in fact, different? Explain your thinking.

You are now going to cut this object along the midline.

9) What do you think will happen?

Go ahead and cut until you return to the starting point.

10) What did you observe?
11) How many sides and edges did you end up with? How did you determine this?

Interlocking loops: Now, you are going to create two interlocking loops. Create one loop (without any twists) and tape the short ends together. Thread another strip of the same size through the first loop and tape the short ends together, again without any twists. You should have two interlocking loops (see Figure 5.9).

Figure 5.9 Two interlocking loops.

12) What do you think would happen if we cut each loop along its midline?

Go ahead and cut each loop along its midline, being careful not to cut any part of the other loop.

13) Hold up your cutout by each of the resulting loops in turn, letting the others hang below. What do you observe?
14) How many sides and edges did you end up with? How did you determine this?

Interlocking Möbius strips: Next, you are going to create two interlocking Möbius strips. To do this, create one Möbius strip. Thread another strip of paper that is the same size through the Möbius strip. Introduce a half-twist to the strip of paper and tape the short ends together. You should have two interlocking Möbius strips (see Figure 5.10).

Figure 5.10 Two interlocking Möbius strips.

15) What do you think would happen if we cut each Möbius strip along its midline? What do you wonder?

Cut each Möbius strip along its midline, being careful not to cut the other Möbius strip.

16) What did you observe?
17) How many sides and edges did you end up with? How did you determine this?

IV. What else?

There is so much more that can be discovered with Möbius strips. Some resulting shapes may be easier to predict than others but they are all equally stunning and mysterious. Let's see what else can happen.

Cutting closer to the edge

Let's go back to the original Möbius strip and cut it closer to the edge. Make a new single Möbius strip. Instead of cutting down the midline, we are going to cut it ⅓ of the way from the edge. This time, imagine two highway dividers dividing the highway into three equal lanes. You are going to cut along one of those dividers.

1) What do you think will happen?

Go ahead and cut. Keep track of what you notice and wonder.

2) How many sides and edges does your object have? How would you explain what happened?
3) What do you think would happen if you cut even closer to the edge? What if you cut ¼ or ⅕ of the way in from an edge?
4) What other variations of half-twists or cuts would you suggest? Describe your variations and jot down your predictions. Try them out and record what you find.
5) In what ways does the Möbius strip represent infinity? What are different ways of measuring the length of a Möbius strip?
6) Consider the poem by William Blake we mentioned earlier. Here are the first four lines. How does it relate to the mathematics you just explored?

"To see a world in a grain of sand
And a heaven in a wild flower,
Hold infinity in the palm of your hand
And eternity in an hour."

V. Story

First formally studied around 1858 by mathematicians August Ferdinand Möbius and Johann Benedict Listing, the Möbius strip remains a source of great interest and wonder today. In addition to being studied in mathematics and science, it appears in literature, art, engineering, computer technology, jewelry design, and company logos.

Where else have you seen the Möbius strip? Think of places beyond this textbook or your mathematics classroom, like local art galleries, sculptures, geological formations, book illustrations, furniture, advertisements, or jewelry. Perhaps you remember coming across a shape similar to the Möbius strip around town or in other places in your daily life.

For example, did you know that Möbius strips can be utilized as conveyer belts? To help you imagine this, think of the moving belt on which you place items for checkout at your local grocery store. Many have a half-twist hidden under the counter. Why do you think it would make sense to use the Möbius strip in this situation? Perhaps you can do a quick search for information about the structure of conveyor belts used in industrial or commercial settings.

The Möbius strip is often studied as an introduction to the field of topology. Topology, a subfield of geometry, is concerned with describing shapes that remain the same (topologically) if you stretch or contort them without cutting or attaching. As opposed to measurements for distance, area, or volume that you undoubtedly learned about in geometry class, topology uses features such as connectedness and existence of holes to describe shapes. For example, a solid cube and ball are topologically the same because neither has a hole and you can imagine morphing one into the other if they were made from malleable material. Two coins, however, can be topologically different. In Figure 5.11, the coin on the left, a Japanese 50-yen coin, and the Polish 5-złoty coin on the right, are topologically different. Even if you could stretch them out, the hole of the Japanese coin distinguishes it topologically from the złoty coin. The złoty coin is topologically the same as a golf ball. Can you think of something that is topologically the same as the 50-yen coin? What about our simple loops (without twists) and Möbius strips (with twists)? Are they the same topologically? In other words, is there a way to stretch or contort a simple loop into a Möbius strip?

Figure 5.11 Two topologically different coins.

VI. Going beyond

1) Create a video to showcase your favorite Möbius strip modification. Share your video with friends.
2) Make art from a Möbius strip.
3) Engineer a contraption that makes use of a Möbius strip. Do you think it is possible to have

    a) A Möbius strip roller coaster?
    b) A Möbius strip on which a ball can roll without falling?

4) What can the Möbius strip represent? What metaphors can the Möbius strip inspire?
5) You are an ad agent. Sell a product using the Möbius strip as a logo.
6) Create an animation that demonstrates your understanding of the properties of a Möbius strip.
7) Research the practical uses of a Möbius strip.
8) Create a fashion design that features a Möbius strip.
9) Maurits Cornelis Escher (usually referred to as M. C. Escher), a 20th-century Dutch artist, is one of many artists who found inspiration in the mathematical properties of the Möbius strip. His illustrations of 'loopy' and interconnected shapes that incorporate features of the Möbius strip invite us to visualize seemingly endless paths and patterns in a new way. Look up some of his sketches and describe where you see Möbius strips (or mathematics more generally) in his work.
10) Find other art that seems to have been inspired by the Möbius strip. Critique it.

VII. Let's reflect

1) Is there such a thing as one-sided paper? Develop an argument for 'both sides' (pun fully intended) of the issue.
2) What surprised you about the Möbius strip?
3) What about this quest made you feel creative?
4) What mathematical processes did you engage in through this exploration?
5) How can your explorations of the Möbius strip help you in other disciplines?

Quest 5.2: Expanding our view of the world through non-Euclidean geometry

What is it like to be a triangle on a sphere?

In this quest, your assumptions about lines, planes, distances, angles, and even common geometric formulas will be challenged. You will study the geometries of shapes like triangles, quadrilaterals, and circles on curved surfaces and test their alignment with the axiomatic foundations of Euclidean geometry. You may even start to look at the world differently!

I. Invitation

Have you ever taken a car ride through an open plain where you could see nothing but the empty fields on the side of the road and horizon ahead of you? What shape was the road? What shape was the horizon? Did anything appear to be completely straight? Now, look around you. From picture frames to books and furniture, there are plenty of objects constructed to be straight and flat. But what shape are trees, plants, and other naturally occurring surfaces around you? Is nature straight and flat?

II. What makes you wonder?

Consider the flight path between New York City, USA, and Tokyo, Japan (see map in Figure 5.12).

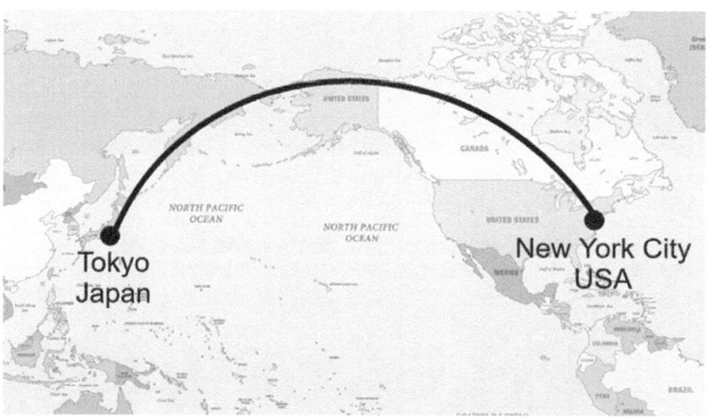

Figure 5.12 A flight path from New York City to Tokyo. World map by Fourleaflover, retrieved from https://www.shutterstock.com/image-vector/world-map-vector-high-detailed-illustration–2480552165. Image modified by the authors.

Whoa, why bother going so far out of the way instead of flying straight across the Pacific Ocean (see the dotted line in Figure 5.13)?

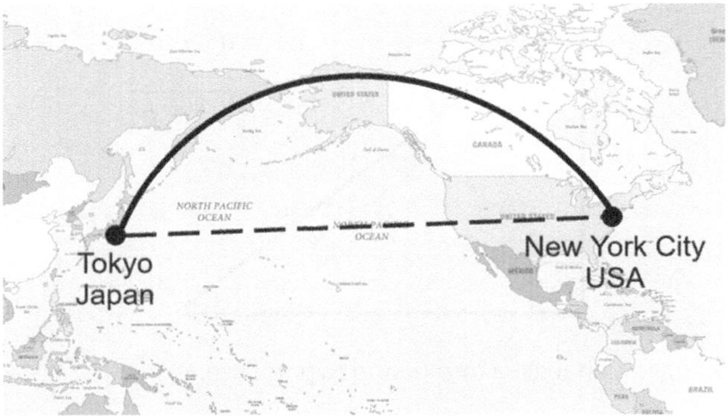

Figure 5.13 An imaginary flight path going 'straight-across' from New York City to Tokyo, denoted by a dotted line contrasted with the presumed actual flight path. World map by Fourleaflover, retrieved from https://www.shutterstock.com/image-vector/world-map-vector-high-detailed-illustration-2480552165. Image modified by authors.

It seems like an egregious waste of time and jet fuel to fly in such a roundabout way. But, could there be a reason? Even a mathematical reason? Let's see what the flight path from Quito, Ecuador, to Nairobi, Kenya, would be (see Figure 5.14).

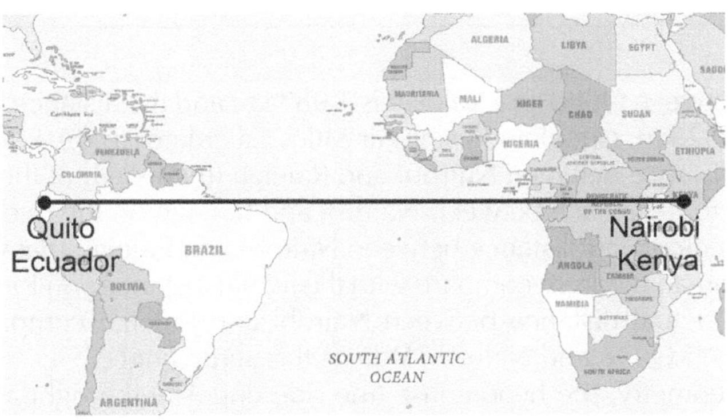

Figure 5.14 A presumed flight path from Quito to Nairobi. World map by Fourleaflover retrieved from https://www.shutterstock.com/image-vector/world-map-vector-high-detailed-illustration–2480552165. Image modified by the authors.

Why do you think the flight path looks curved for New York City to Tokyo but looks straight for Quito to Nairobi?

To investigate this further, recall the Pythagorean theorem. This theorem states that given a right triangle, the square of the length of the hypotenuse (the side opposite the right angle), c, is equal to the sum of the squares of the two legs (the other two sides), a and b. That is, $c^2 = a^2 + b^2$. It follows that the length of the hypotenuse is the square root of the sum of the squares of the legs (see Figure 5.15). That is, $c = \sqrt{a^2 + b^2}$.

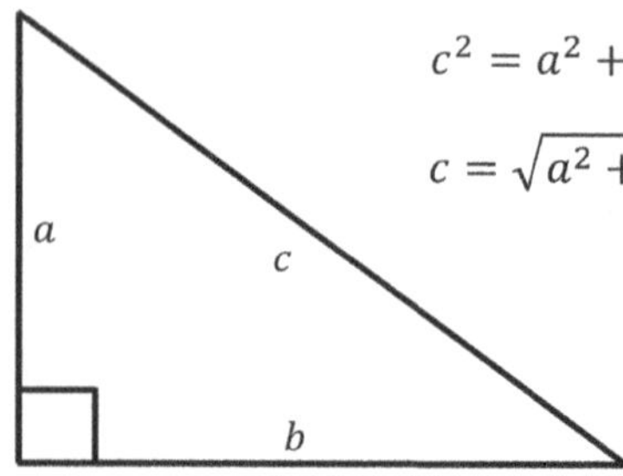

$$c^2 = a^2 + b^2$$

$$c = \sqrt{a^2 + b^2}$$

Figure 5.15 A right triangle with legs, *a* and *b*, and hypotenuse, *c*.

We are going to use the Pythagorean theorem to calculate the distance between Nairobi, Kenya, and Raleigh, USA. Look up the location of Quito, Nairobi, and Raleigh on a map. Notice that they can be seen to represent vertices of a right triangle. Draw this right triangle. Copy Table 5.1 and use it to record your answers to the following questions.

Table 5.1 Table for recording calculated and actual distances between Nairobi and Raleigh.

Actual distance between Quito and Raleigh	*Actual distance between Quito and Nairobi*	*Calculated distance between Nairobi and Raleigh*	*Actual distance between Nairobi and Raleigh*
3,987 km	12,832 km		

1) The distance between Quito and Raleigh is 3,987 km and the distance between Quito and Nairobi is 12,832 km, calculated using Air Miles Calculator (n.d.). Use these distances to calculate the distance between Nairobi and Raleigh (the length of the hypotenuse).
2) Look up the actual distance between Nairobi and Raleigh on the Internet.
3) How does the calculated distance between Nairobi and Raleigh that you determined using the Pythagorean theorem compare with the actual distance you found on the Internet?
4) How does the actual distance between Nairobi and Raleigh compare with the actual distance between Quito and Nairobi? Why is this surprising?
5) In Euclidean geometry, the hypotenuse (the side opposite the right angle) is the longest of the three sides of a right triangle. Why do you think this property does not hold when you consider the actual distances between these cities?

Let's see if we can figure out what's going on. Take a moment to prepare for the activity:

- Find a beachball. The bigger the better! Blow it up.
- You will need a piece of string or large rubber bands long enough to go around the beachball. If you don't have access to string or large rubber bands, similar materials such as dental floss, yarn, or tape can also work well.
- If you have an inflatable globe (which perhaps your instructor has handed you), let's have that handy, too.
- Print and cut out a paper ruler and a paper protractor. These need to be flexible.
- Find an erasable (water-based) marker and tape.

6) What shape is your beachball?

Place the tip of your finger on the beachball. Trace a path from that spot to any other spot on the beachball. As you do this, look closely at the shape of the path you are tracing (see Figure 5.16).

Figure 5.16 An example of a path drawn on a beachball.

The geometry of shapes on the surface of a sphere – like on Earth or on our beachball, both of which are nearly spherical – is called **spherical geometry**. Note that 'lines' in spherical geometry are curved because they follow the curved surface of the sphere – the same way that the path you traced on the beachball followed the curve of the beachball. This is in contrast to the geometry you are most likely familiar with, called **Euclidean geometry,** which describes geometry on a (flat) plane.

Now take your string (or similar material) and wrap it around the beachball so that you have created a border between two halves of the sphere. If your beachball is like the one in Figure 5.16, you can trace the existing 'lines' on the ball to create two halves of a sphere, but try to experiment with placing the string at different orientations as well. Do this multiple times. In each case, make sure that your string 'slices' the beachball in half.

7) Can you see that there are an infinite number of places where you can wrap your string such that it serves as a border between two halves of the sphere?

You have just created **great circles**. A great circle is an arc on a sphere that ends at its starting point and creates a circle that shares its center with the center of the sphere. It also 'slices' the sphere into two halves. In Euclidean geometry, the shortest distance between two points on the plane is the length of the line segment connecting the points. In spherical geometry, the shortest path between any two points lies along a great circle (see Figure 5.17).

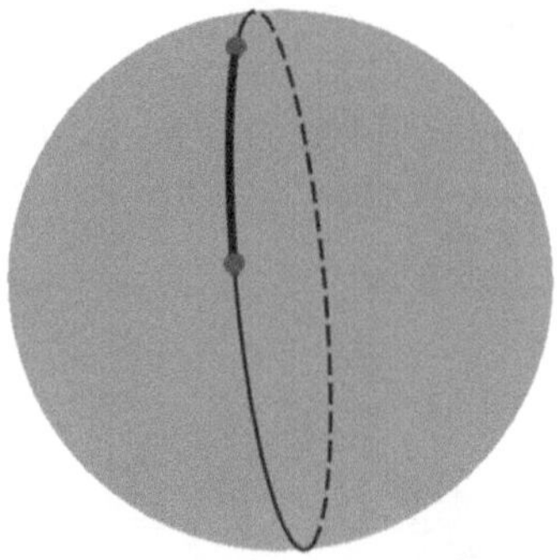

Figure 5.17 A great circle denoted by a solid line (front) and a dashed line (back). An arc that lies along the great circle is shown in bold. Image constructed by Preston Pietruszewski using Maple 2024.1.

Let's keep looking for more special characteristics of great circles.

8) Great circles 'slice' the sphere exactly in half. What else do great circles have in common? Use the beachball to find as many examples of commonalities as you can.
9) Great circles are not the only circles that can be drawn on the surface of a sphere. What other circles are possible?

Imagine a plane slicing a sphere into two *unequal* pieces.

10) Does the plane (and the resulting slice) pass through the center of the sphere?
11) What does a slice or cross-section look like? Try drawing several possibilities.

When you 'slice' a sphere to create two unequal pieces, the circle that is the intersection of the 'slice' and the sphere is called a **small circle**.

12) What do all small circles have in common? How are they different from great circles?
13) Is every circle drawn on a sphere that is not a great circle a small circle?
14) Earth has two geographic poles, the North Pole and the South Pole. These poles are the points where the axis of rotation of Earth and the surface of Earth intersect. Where on a globe do you see great circles that go through these two poles? How many are there? What are these called?

The equator is the only circle with an east-west orientation that is a great circle. All other east-west circles are small circles which are parallel to the great circle that is the equator. Verify this fact on your globe. Okay, now back to that flight path ... .

Now that you know a bit more about spherical geometry, let's develop a conjecture as to why the shortest flight path from New York City to Tokyo doesn't go 'straight' across North America and the Pacific Ocean.

Mark two points on the surface of your beachball, inflatable globe, or another large spherical object that approximate the locations of New York City and Tokyo. Using string and tape, create two possible paths between the points:

- A path along a great circle, and
- A path along a small circle.

15) What do you notice about the paths? Which path is shorter? How did you determine this?

On Earth, the shortest distance from one place to another lies along a great circle. This property of spherical geometry is used in navigation as it helps to minimize the monetary and environmental costs of travel. For instance, our example of the flight path from New York City to Tokyo goes over Canada, Alaska, and the Kamchatka Peninsula and is about 10,899 km, calculated using Air Miles Calculator (n.d.).

Use the Internet or another method to determine the distance between New York City and Tokyo that roughly follows the latitude line that goes through New York City (the path along the dotted line on the map in Figure 5.13).

16) How does this distance compare to the distance along the great circle? How can you use your new knowledge about spherical geometry to explain this result?

Find two points on your beachball that approximate the locations of Quito, Ecuador, and Nairobi, Kenya. Use your beachball, string, ruler, and tape to investigate the distance between these points (cities) using different paths. You can either use a paper ruler or use string and a standard ruler to approximate these measurements.

17) Which path appears to be the shortest? Why?
18) What are you wondering about now?
19) We asked you to create a right triangle using Quito-Nairobi as one leg and Quito–Raleigh as another to calculate the distance between Raleigh and Nairobi earlier. We found that calculating this using the Pythagorean theorem led to a distance that is much larger than the actual distance.
    a) How would you explain this now?
    b) Why do you think we chose these cities for our example?
    c) What other group of three cities could we have chosen for a similar problem?

Next, you will explore what other well-known mathematical 'truths' do not hold in spherical geometry.

III. Let's see what happens

The three angles on the inside of a triangle, the **interior angles**, add up to 180°, right? Is this always true? Can you imagine it not being true?

1) Are there times in mathematics when 'it just depends'?

For this next activity, you will need a *paper* protractor. It must be flexible since we will be using it to measure angles on surfaces that are not flat. It may have been a while since you used a protractor, so let's review how it works.

Here is an angle (see Figure 5.18):

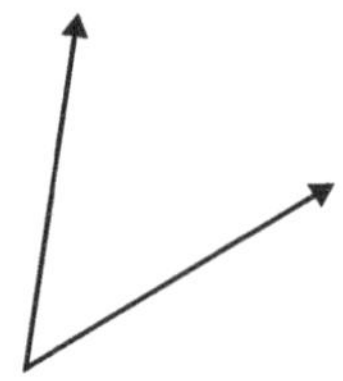

Figure 5.18 An image of an angle.

2) Do you think this angle is greater than or less than 90°? How do you know?

Line up the vertex of the upside-down 'T' of the protractor with the vertex of the angle, as shown in Figure 5.19.

Find '0' (or 0°) on the protractor and count up until the tick mark on the protractor and the other ray of your angle line up. You may need to extend the rays of the angle to do this. Be sure to count up from 0° and use the 'inner' tick marks (instead of counting down from 180° using the 'outer' tick marks). Confirm that the angle shown is about 50°. If you read 130°, you most likely read the wrong set of tick marks.

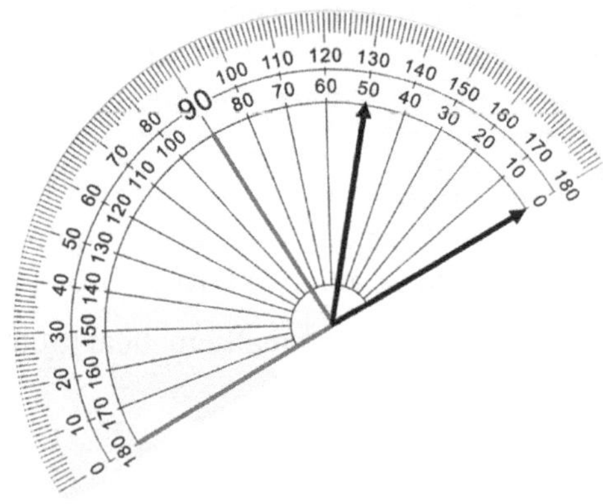

Figure 5.19 Measuring a 50° angle with a protractor. Image modified by the authors.

Use your protractor to confirm that the sum of the interior angles of each triangle in Figure 5.20 is 180°.

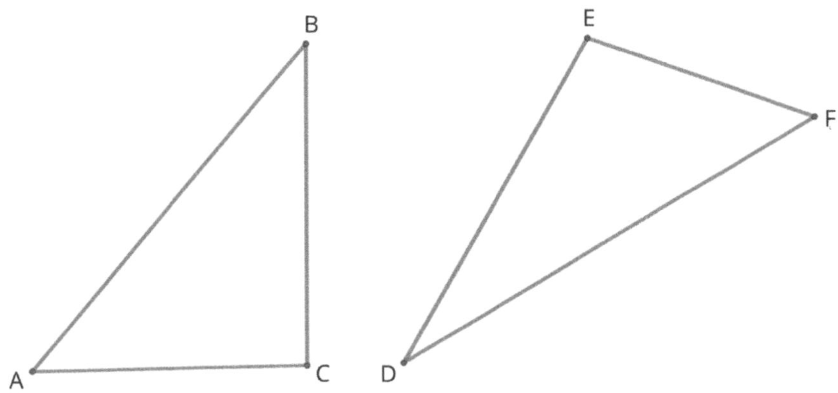

Figure 5.20 Two triangles (ABC and DEF) for calculating the sum of the interior angles.

Find other triangles around you on flat surfaces and measure their angles. The sum of the measures of the interior angles should be 180° for each triangle.

Now back to that beachball … . Use a marker to draw at least three triangles of different sizes on your beachball. To make each triangle, start by drawing three points on the sphere and then connecting each pair of points using the shortest route (which will be along a great circle – see Figure 5.21) either by drawing with your marker or by taping string to the sphere.

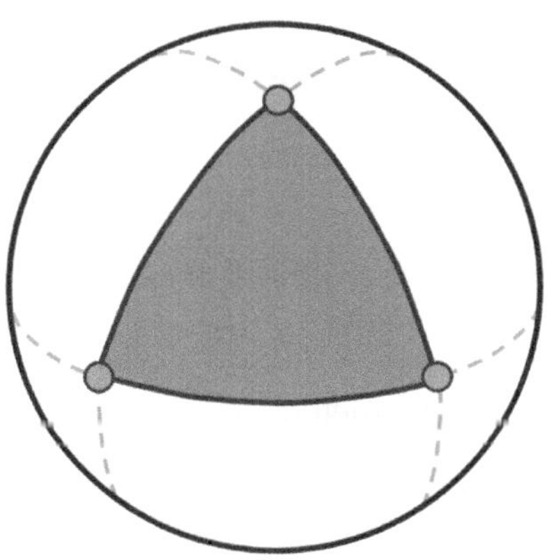

Figure 5.21 A triangle on the surface of a sphere. Image constructed by Preston Pietruszewski.

3) What differences do you notice between triangles on a flat surface and triangles on a sphere? To what can you attribute these differences?
4) What differences do you notice between the smaller and larger triangles that you drew on your beachball? To what can you attribute these differences?
5) Use your paper protractor to measure the angles of each triangle on your beachball. Find the sum of the angles for each of your triangles. What do you wonder about now?
6) What are some differences between the angles of triangles drawn on a sphere and those drawn on a flat surface?
7) What was the largest angle sum you found among the triangles you drew on your beachball? Can you draw a triangle that has an even greater angle sum?

IV. What else?

You should have found that the sum of the interior angles of a triangle drawn on a sphere exceeds 180°. This intriguing finding runs counter to our understanding of Euclidean geometry.

1) What do you think is the largest sum of interior angles of a triangle on a beachball? Why do you think this?
2) What assumptions are you making about the triangle?

Triangles on a sphere

Draw a great circle on your beachball (see Figure 5.22) and then draw three points on your great circle.

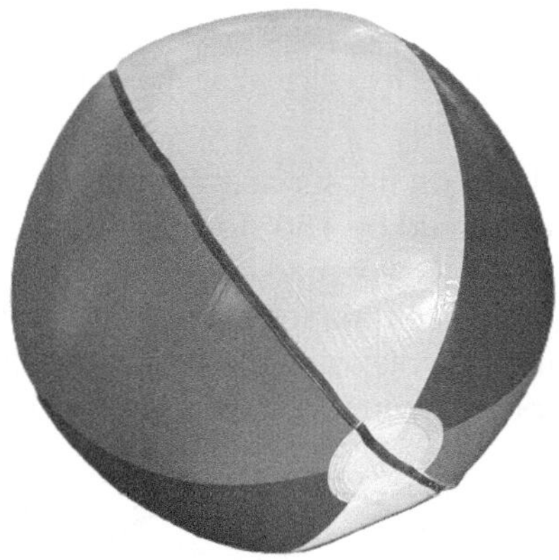

Figure 5.22 Great circle on a beachball.

3) Could these three points be the three vertices of a triangle? Why or why not? If yes, what is the sum of the angles of this triangle?
4) How does your response to Question 3 influence your hunch about the largest sum of the angles of a triangle in spherical geometry?

We might also argue that the sum of the angles of a triangle drawn on a sphere can be up to 900°.

5) Why is that?

Let's reason through this using Figures 5.23 and 5.24. First, consider Figure 5.23. To help us visualize interior and exterior angles, we have drawn a circle around the vertex and used shading to distinguish between the two types of angles. Figure 5.24 shows a triangle with three interior angles (see the shaded portion of each circle) and three exterior angles (see the unshaded portion of each circle).

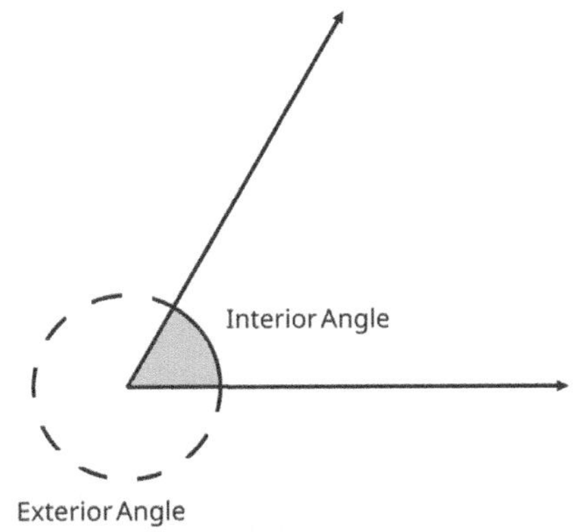

Figure 5.23 An example of an interior angle and its corresponding exterior angle.

6) What is the sum of the measures of an interior angle and its corresponding exterior angle?

Now consider the Euclidean triangle in Figure 5.24.

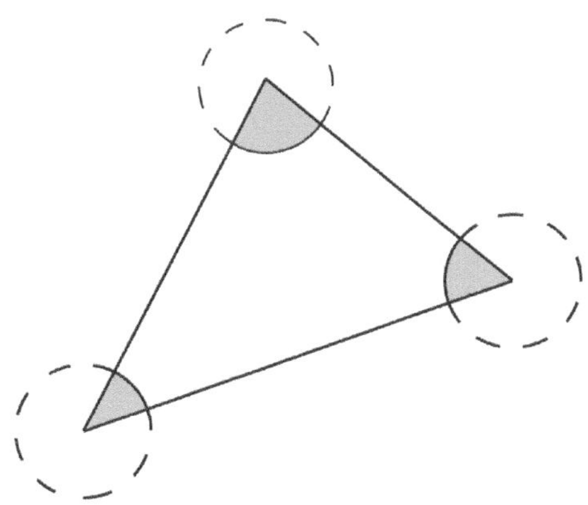

Figure 5.24 Euclidean triangle showing three interior angles (see the shaded portion of each circle) and three exterior angles (see the unshaded portion of each circle).

7) What is the sum of the exterior angles of any Euclidean triangle? How do you know this?

Now we are going to consider how the sum of the exterior angles of a triangle on a flat surface might be used to argue why the sum of the interior angles of a triangle drawn on a sphere can be up to (but not equal to) 900°.

If we define a triangle to be a polygon with three vertices bounded by the three edges that connect them, we must consider what this can look like on a sphere. Note that when a triangle is drawn on a sphere (see Figure 5.25), its three edges actually define two triangles instead of one. There are two regions bounded by the three sides. In Figure 5.25, they appear with two different shadings. Together, these two triangles cover the entirety of the surface of the sphere.

Figure 5.25 Two triangles (one black, the other gray) on a sphere. Image constructed by Preston Pietruszewski.

For our purposes, we define the interior triangle as the one that encloses the smaller surface area. It is also the triangle that subjectively looks more like the triangles that we are familiar with. The interior of the triangle is curved, just like the surface of the sphere. The exterior triangle probably doesn't look like any triangle you've seen before, and may be a bit harder to visualize. Do you see why it might be considered a triangle?

Use what you know about the sum of all of the interior and exterior angles of a triangle and the fact that there are two possible triangles bounded by three edges (one interior and the other exterior) to convince yourself that the sum of the angles of triangles drawn on the surface of a sphere can be up to 900°.

Quadrilaterals on a sphere

Let's take this argument beyond triangles. What do you think are the smallest and largest sums of the angles of a quadrilateral drawn on the surface of a sphere? Imagine such a quadrilateral. Make a prediction. Figure 5.26 shows four interior and four exterior angles of a quadrilateral drawn on a plane. Use it to construct your argument.

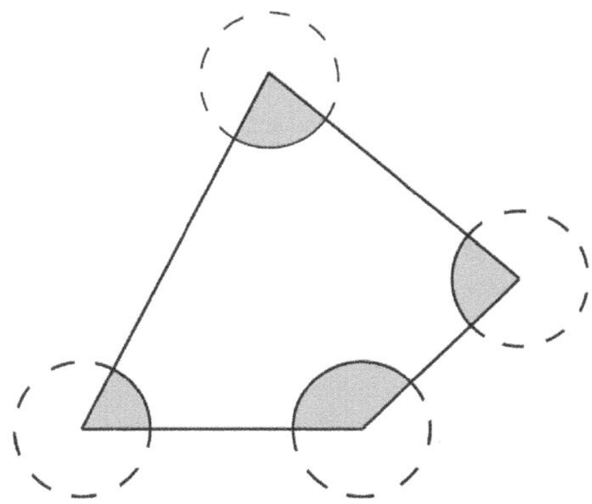

Figure 5.26 A quadrilateral on a plane showing four interior angles located on the shaded side and four exterior angles located on the unshaded side.

Continue investigating by drawing a quadrilateral on the surface of your beachball. Measure the angles with a paper protractor. Figure 5.27 shows a quadrilateral drawn on the surface of a sphere.

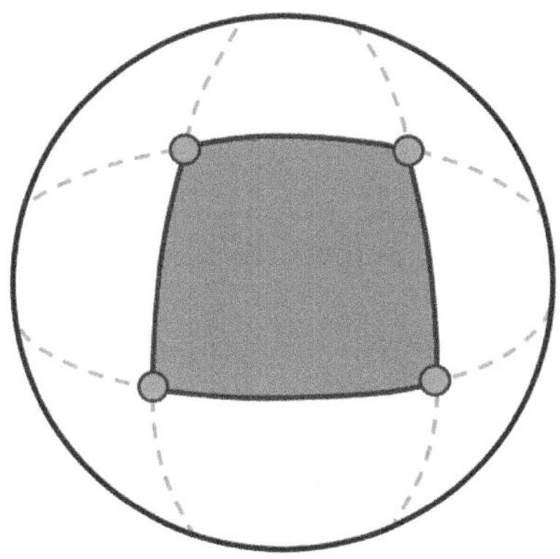

Figure 5.27 A quadrilateral on the surface of a sphere. Image constructed by Preston Pietruszewski.

Use Figures 5.26 and 5.27 and our arguments from the example of a triangle in spherical geometry to answer the following questions:

8) What is the smallest possible sum of the angles of a quadrilateral drawn on a sphere?
9) What is the largest possible sum of the angles of a quadrilateral drawn on a sphere?

Polygons on a sphere

Take this exploration one step further and consider pentagons (five-sided polygons) on a sphere.

10) What are the smallest and largest possible sums of the angles of a pentagon drawn on a sphere? Explain your reasoning.

Generalize this conjecture for any *n*-gon (a polygon with *n* sides).

Triangles on other surfaces

You have explored how triangles drawn on a sphere are different from triangles drawn on a plane. Let's now consider what happens when we draw triangles on other surfaces. The shape you see in Figure 5.28 is a 'saddle'. It might remind you of a horse saddle or a Pringles® potato chip. Geometry on a saddle is known as **saddle geometry**. The saddle is one of a few possible representations of **hyperbolic geometry**.

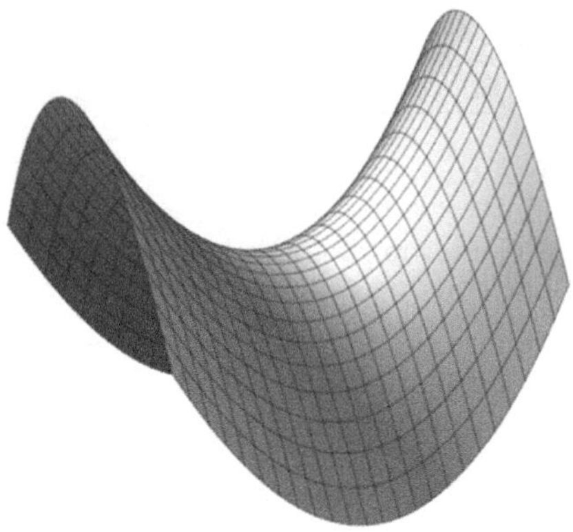

Figure 5.28 A 'saddle' shape in hyperbolic geometry. Image constructed by Preston Pietruszewski using Maple 2024.1.

If you can, get a hold of a saddle-shaped object, like a Pringles® potato chip. If you can't find one, look up a saddle shape on the Internet. Consider the following:

11) In what ways might a triangle drawn on a saddle differ from our previously explored triangles?
12) Draw several triangles on your saddle. If you are using a Pringles® chip, find a soft, felt-tipped marker so it doesn't crumble. Find the sum of interior angles of triangles drawn on a saddle. Describe each triangle and note the sum of its interior angles.
13) Based on your explorations of geometry so far, what can you infer about the properties of triangles on a saddle? Why do you think this is?

V. Story

You have taken the first steps into the world of non-Euclidean geometry, where the properties of shapes, angles, and lines differ from those in Euclidean geometry. Most geometry lessons in school focus on the principles of Euclidean geometry, set out around 300 BC in *The Elements* by the Greek mathematician Euclid. This text remains one of the most studied collections of postulates, definitions, and propositions today.

Do you remember learning definitions, theorems, and formulas in geometry class? Among them, there were postulates that you did not have to prove but rather accepted as 'obvious'. You might recall the five postulates from Euclid's *Elements*.

The following are a translation of the original postulates (Euclid, 300 BC/2006). You may need to read each more than once. Doodle while you are reading the postulates. See if you can make sense of each as you doodle:

Postulate 1: To draw a straight line from any point to any point.
Postulate 2: To produce a finite straight line continuously in a straight line.
Postulate 3: To describe a circle with any center and distance.
Postulate 4: That all right angles equal one another.
Postulate 5: That, if a straight line falling on two straight lines make(s) the interior angles on the same side less than two right angles, the two straight lines, if produced indefinitely, meet on that side on which are the angles less than the two right angles.

Turn the page after you have finished doodling to see other interpretations of these postulates.

1) Did you doodle something similar to the drawings in Figure 5.29? If not, how were your interpretations of the postulates different from those shown in Figure 5.29?

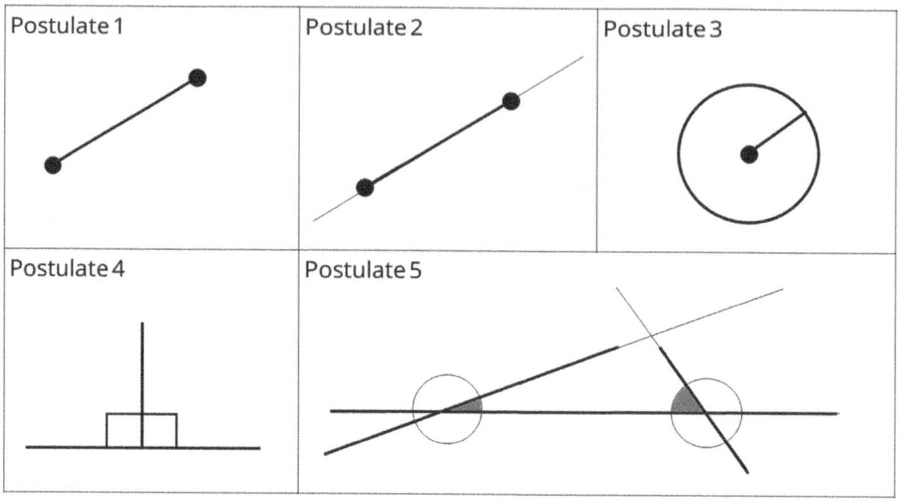

Figure 5.29 Visual representations of Euclid's five postulates.

2) What do Euclid's postulates mean? Rewrite each in your own words.

Answer these questions about the implications of each of the five postulates:

3) How many line segments can be drawn between two points?
4) How can lines be constructed?
5) What defines a circle?
6) What makes all right angles congruent to each other? What would a square look like if this wasn't true?
7) Do parallel lines ever meet?

Euclid's five postulates helped build a system of hundreds of interconnected theorems.

8) What would happen if one or more of the five original postulates did not hold?

We will focus on Postulate 5 to explore this question. Let's go back to the beachball. Wrap a string around the ball to make a great circle (see Figure 5.22). Choose a point not on the great circle.

Euclid's Postulate 5 can be rephrased in the following way (see the 'Going beyond' Question 4 for an exploration of why this is):

> Given a point not on a straight line,
> there exists one and only one straight line
> through the point parallel to the given straight line.

Let's replace the words 'straight line' with 'great circle' to see if Postulate 5 holds in spherical geometry. Recall that great circles are analogous to lines when working on a sphere.

> Given a point not on a great circle,
> there exists one and only one great circle
> through the point parallel to the given great circle.

Test this modified Postulate 5 on your beachball.

9) What about Postulate 5 breaks down in spherical geometry?

By revisiting Euclid's postulates, we challenged our presuppositions about mathematics and built new understandings about the world around us.

VI. Going beyond

1) Consider the following three axioms:
    - There are five flavors of ice cream: vanilla, chocolate, strawberry, bubble gum, and cookie dough.
    - Given any two flavors, there is one child who likes both of these two flavors.
    - Every child likes exactly two different flavors among the five.

    a) Use these axioms to determine how many children there are. Show the logic behind your work and also show that your result satisfies each of the axioms.
    b) In your own words, define the word 'axiom' based on your work in part (a).
2) Write a set of your own axioms and ask a friend to derive a new theorem based on your axioms.
3) Find other types of curved or uneven surfaces. Repeat similar calculations to those done in the quest on these surfaces and consider how the results change. Look for patterns and connections between shapes and their many features.
4) The following are two equivalent versions of Euclid's Postulate 5:

    Original version (Euclid's *Elements*): That, if a straight line falling on two straight lines make(s) the interior angles on the same side less than two right angles, the two straight lines, if produced indefinitely, meet on that side on which are the angles less than the two right angles.

    Rephrased version: Given a point not on a straight line, there exists one and only one straight line through the point parallel to the given straight line.

    Explain why these two versions can be used interchangeably.

VII. Let's reflect

1) In this quest, you 'sliced' a sphere to form circular cross-sections. Explore slicing a bit more. Can non-circular shapes be created with a single slice of a sphere? What shapes could result from two slices or more?

2) What shapes result when you slice a cube in different ways? How many different shapes are possible? What about slicing a cone? A square pyramid?
3) What does geometry mean to you now? Write a letter to your former geometry teacher about your discoveries in this lesson.
4) What surprised you about non-Euclidean geometry? Why was this surprising?
5) How else (besides influencing flight paths) might this quest be pertinent in your daily life?

Quest 5.3: Examining what we value through mapmaking

Is there a perfect map?

In this quest, you will explore **cartography,** the science and art of mapmaking. Certain decisions must be made when representing a curved surface, like Earth, on a flat surface, like a map. Cartography is rich with innovative and purposeful designs used to improve the accuracy of these maps. In this quest, you will consider how maps can reveal different information depending on how they represent Earth. We would like for this last quest to serve as both a reminder of some of the themes you have explored throughout your journey in this book and an invitation for you to carry your imagination, innovation, and creative thinking into your lives beyond the scope of this book. We hope you never stop seizing opportunities to be creative in your future studies and life experiences!

I. Invitation

Have you ever spun a globe? Perhaps you own an actual, physical globe. Perhaps you have spun a globe virtually using an interactive map application. Spinning a globe allows us to see parts of Earth, a little at a time. What happens when we try to view the entire Earth on a flat map so that we can see all of it at once?

In our day-to-day lives, we are more likely to look for locations on map applications on a computer, tablet, or phone than on a globe. Or, perhaps you have an old-fashioned paper map or atlas handy. All of these maps depict Earth on flat surfaces. Recall that in Quest 5.2, we studied the implications of thinking of Earth as a flat surface when it is actually (almost) a sphere.

World maps are two-dimensional representations of Earth and its features. They can show physical and artificial boundaries between countries, natural resources, weather patterns, landscapes, bodies of water, or climate differences. How these maps represent Earth is important: Depending on what features of Earth are emphasized, maps have the potential to influence significant decisions and policies.

Open an interactive map application on an electronic device and find your location. Adjust your screen so that you can also see a place that is 1000 km away from you.

1) What location did you choose?
2) What if you zoom out even more? Can you zoom out enough to see the whole world?
3) Is there a limit to how much you can zoom out? What eventually happens?
4) What do you notice? What do you wonder?
5) Use a different application to see if there are any differences in what happens when you zoom out.

II. What makes you wonder?

We rely on the accuracy of maps nearly every day to go about our daily lives. Maps make it easier for us to move from place to place, avoid traffic jams, and lower the cost of travel. New maps are designed and older ones are revised all the time so that we can have access to more accurate and useful maps.

1) Examine the image of a world map shown in Figure 5.30. What are its key features? What do you wonder about?

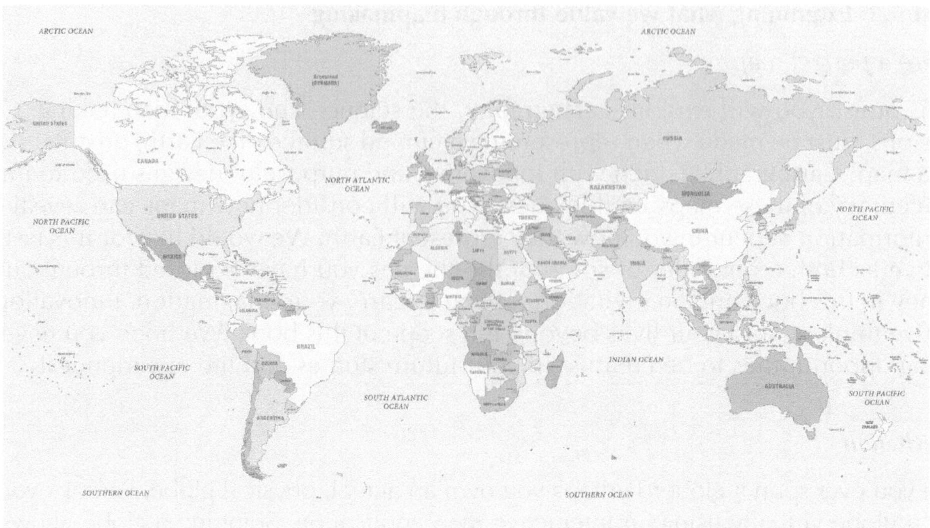

Figure 5.30 A map of the world. Image by Fourleaflover, retrieved from https://www.shutterstock.com/image-vector/world-map-vector-high-detailed-illustration–2480552165.

2) What important features of Earth or a globe seem to have been prioritized in the map shown in Figure 5.30?
3) Is anything missing from this map? Are there parts of Earth that are misrepresented?
4) According to Wikipedia, the area of India is approximately 3.29 million km^2. The area of Greenland is approximately 2.17 million km^2.

   a) Based on this information, calculate the ratio of the area of India to the area of Greenland. How many times larger is India than Greenland.
   b) Consider the map in Figure 5.30 again. Use the map to estimate the ratio of the area of India to Greenland. Be sure to describe how you arrived at your answer. Thinking about the kind of reasoning you used to estimate in Quest 1.2 and how you used the Monte Carlo method in Quests 3.2 and 3.3 may help you with this problem.
   c) How does the ratio from a) compare to the ratio you found in b)?
   d) What do you think is the cause of this discrepancy?

5) Find Kenya and Ecuador on the map in Figure 5.30.

   a) Use the map to estimate the ratio of the area of Kenya to the area of Ecuador. Be sure to describe how you arrived at your answer.
   b) Find the area approximations of Kenya and Ecuador on the Internet. Calculate the ratio of the area of Kenya to the area of Ecuador.
   c) How does your ratio from a) compare to the ratio you found in b)?
   d) You've seen Quito, Ecuador, and Nairobi, Kenya, before in this book (in Quest 5.2). What features do these cities share that is relevant to this quest? Why do you think we keep bringing up these cities?

6) Which countries are the most distorted on the map in Figure 5.30? Why do you think this is?

7) Recall that in Quest 3.2, you tossed an inflatable globe to estimate the percentage of Earth's surface that is covered by water using the Monte Carlo method.
   a) What was your estimated percentage using the Monte Carlo method in Quest 3.2? (If you don't have your solutions handy, estimate this percentage by looking at a globe.)
   b) What is your estimate of this percentage based on the map in Figure 5.30?
   c) How would you explain this discrepancy?
8) According to Wikipedia, the surface area of Earth is approximately 510 million km^2.
   a) Use this fact and the map in Figure 5.30 to estimate the area of the Earth's surface that the Pacific Ocean covers.
   b) Recall that in Quest 1.2, you attempted to determine an estimate for the area of the Pacific Ocean within an order of magnitude of the actual value. (If you don't have your solutions handy, estimate this percentage by looking at a globe.) How does that value compare to the estimate you found in Question 8a?
   c) Which is more accurate – the estimate you developed without a visual prompt (in Quest 1.2) or the one that resulted from considering the map in Figure 5.30? Why do you think this is?
9) Proportional reasoning can be used to identify and convey distortions of a map. In what other quests was proportional reasoning used to communicate distortions or discrepancies?

III. Let's see what happens

Figure 5.31 shows 10 world maps that represent different projections of Earth onto a flat surface. **Map projections** are the result of **projecting** – or transforming – locations on the curved surface of Earth onto a flat surface. Pay attention to the overall shape of each map and the shapes and sizes of different land masses and bodies of water represented on it. Which map is your favorite? Why?

Choose three projections from Figure 5.31 that are significantly different from one another. Answer the following questions for the three map projections you chose:

1) What are the advantages of each map?
2) What are the limitations of each map?
3) What do the three maps have in common?

Use the language and tools you developed in the 'What makes you wonder' section of this quest to answer the following questions about the three maps you chose:

4) How does each map distort Earth?
5) What feature(s) of Earth does each map prioritize?
6) How do you think each map reflects what the mapmaker valued?
7) Which of the three maps do you think is the most accurate projection of Earth? Why?

Some of the maps in Figure 5.31 are more recent than others. For example, while many of these maps were created in the 20th century, the Mercator projection was developed in the 16th century and the origins of the Nicolosi globular projection can be traced to the year 1000. Each map originated with someone's new idea, leading them to 'project' their vision of the world. How do *you* see the globe projected onto a flat surface?

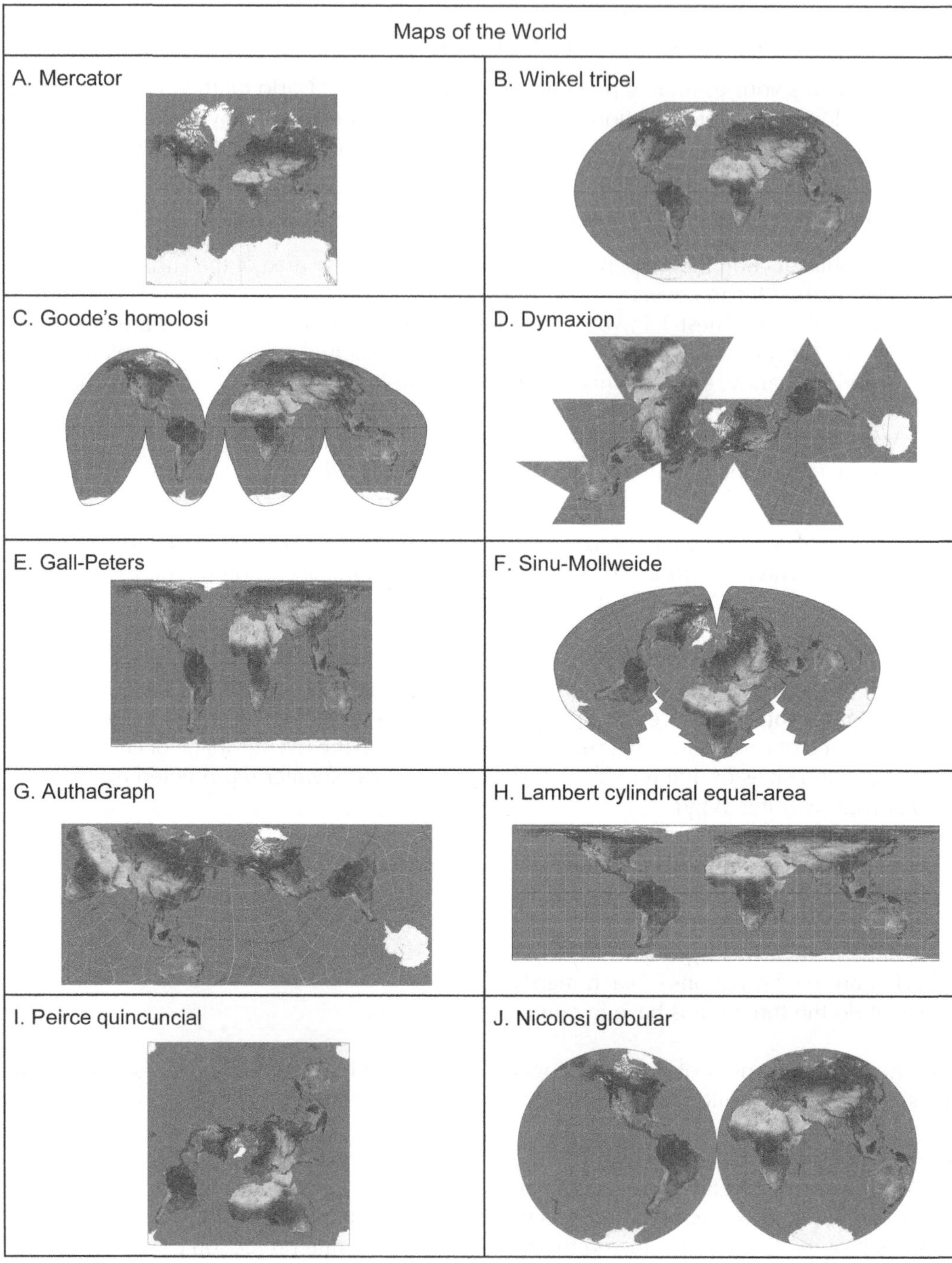

Figure 5.31 Ten map projections of the world.

Images by Daniel R. Strebe (Maps A, B, C, E, F, H, I, and J), Justin Kunimune (Map D), and Felagoth (Map G), retrieved from https://en.wikipedia.org/wiki/List_of_map_projections.

IV. What else?

You have been hired to design a world map that depicts all of Earth. What purpose would you like your map to serve? You decide.

Use the following prompts to guide your thinking:

- What do I value in a map?
- How should I design my map to make it accurate?
- What features will make my map accurate?
- What types of confusion will I try to avoid?
- What resources and tools will I need to design my map?

Your challenge is to create an original map of the world using a method and materials of your choice and to defend the design of your map. It can be one- or two-sided and can be on one or more pieces of paper. Be ready to use mathematical language to argue that your map accurately reflects all of Earth. Be creative in selecting the materials for your design.

Answer the following questions after you've created your map:

1) What are your map's key features? What makes it unique?
2) How would you convince someone that your map is accurate? What calculations would you use to do this?
3) Recall that in Quest 5.2, we saw the discrepancies that arise when comparing distances on the surface of Earth (a curved surface) to those on a map (a flat surface).
   a) Recall that you calculated the distance between Nairobi and Raleigh using the Pythagorean theorem. You found that this calculated distance was different from the actual distance of 12,397 km. Determine a way to estimate the distance between these two cities on your map and compare it to the actual distance.
   b) Choose three cities on your map that can be used as vertices of a right triangle. Draw a triangle with these cities as vertices. Find the area of your triangle. Find these same cities on a map application that can calculate the area of selected regions on Earth (such as Google Earth) and use the application to determine the area of the triangle defined by the same three cities. Do calculations using your map lead to a similar area for this region? How did you determine this?
4) Compare your map to a globe.
   a) What are some key discrepancies?
   b) Is any feature of Earth missing on your map? Is the area of any region exaggerated or diminished relative to the actual area?
   c) What could you have done to minimize these distortions? What would have been compromised in doing so?
5) Describe the constraints you faced when making your map.
   a) How did you address them?
   b) In what way did the constraints help you be creative?
   c) In what other quests did constraints encourage your creativity in mathematics?

V. Story

You may be wondering what is lost when we 'flatten' Earth. Is anything gained? A projection is a mapping of points from one space to another. In our quest, we are projecting locations on the curved surface of Earth onto a flat surface, like a map. What other projections are you familiar with?

Shadows can be quite beautiful to look at. Have you ever wondered what makes a shadow look the way it does? Next time you are on a walk in the park, pay close attention to shadows cast by trees, clouds, or lamp posts. Shadows are formed when an object blocks a light source, such as a light bulb or the Sun. Shadows can be understood as being the result of projections of objects onto the ground or other surfaces. Just like maps, shadows can distort the shape and size of objects.

Earlier in this quest, you saw how attempts to project Earth's curved surface onto a flat map result in distortions in distance, area, or geographic features. A map projection is a specific **transformation**; that is, it represents a rule that maps every point (pre-image) of an object to another point (image). What can mathematics teach us about this? Do all transformations distort distance?

Do you remember reflections (flips), translations (slides), and rotations (turns) from past geometry classes? These special transformations preserve both the shape and size of the object, but not all transformations behave in this way. For example, **dilations** and **circle inversions** can change – or transform – the size or shape of objects. Let's take a closer look at each of these transformations.

A dilation is a transformation that enlarges or shrinks a shape by a scale factor relative to a fixed point, *P*. Imagine doubling the length of the sides of a triangle (see Figure 5.32).

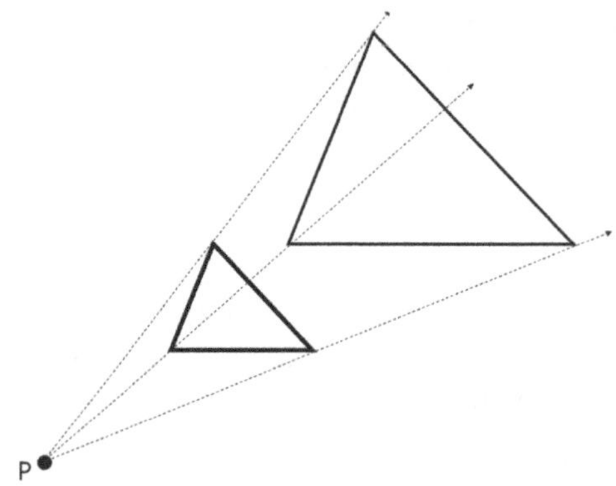

Figure 5.32 Dilation of triangle about a point. (This shows the smaller triangle being dilated about point *P*.)

This is a dilation with a scale factor of 2 about a fixed point, *P*, called the center of dilation. What is different between the two triangles in the figure? What is still the same? What would have happened if the fixed point was in a different location with respect to the original triangle?

Figure 5.33 is an example of inversion with respect to a circle. Circle inversion is a transformation that maps points outside of a circle to points inside of that circle (and vice versa) according to a specific formula – with points on the circle being mapped to themselves. Both the size and shape of objects inverted in circles can change considerably. This figure shows points on line ℓ being inverted with respect to the larger circle that is centered around point *P* to produce the smaller circle centered around point *C*. It shows that lines inverted with respect to circles become circles. Search the Internet for more information about objects inverted in a circle. Be ready to be surprised!

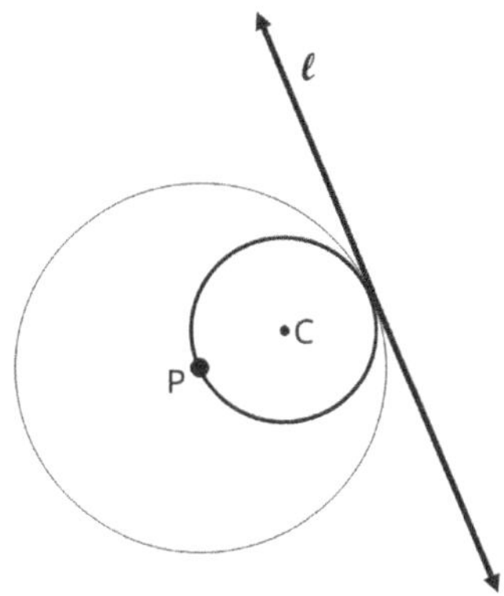

Figure 5.33 Points on line ℓ being inverted with respect to the larger circle that is centered around point *P* to produce the smaller circle centered around point *C*.

Now, back to mapmaking ... . What is mathematically possible? What transformations should be used to preserve distance on a map? World maps continue to be improved today as new ways to project Earth onto flat surfaces are being explored. In 2021, J. Richard Gott, Dave Goldberg, and Bob Vanderbei published a new map in their *Scientific American* article "The Most Accurate Flat Map of Earth Yet" (Gott, 2021). This article presented their two-sided circular map that is more accurate than the Winkel tripel, the map they had previously found to be more accurate than any other map. Their new map aimed to minimize distortions in area and shapes, continuing to push the boundaries of modern mapmaking.

VI. Going beyond

1) Search the web for information about a map design not mentioned in this quest. Describe the design. What made it interesting to you? What, if anything, did you find surprising about the design? Was anything about it controversial? What impressed you?
2) What else can you create a map of? Can you make a map of your room? What would it look like? Would it have the same limitations as map projections of Earth?
3) Find a sculpture or other interesting 3D object. Represent it on paper (or another flat surface) and convince others that your drawing accurately depicts the sculpture.
4) Think of the challenges (and advancements) in mapmaking that you have encountered in this quest, including the experience of making your own map. Make a video in which you share your understanding of the quality of today's maps. Share your video with friends and family or on social media.
5) Read the article, "The Most Accurate Flat Map of Earth Yet" by J. Richard Gott. The URL for this article appears in the 'Quest 5.3: Cited and related works' section at the end of this chapter.

a) Reflect on what you read by answering the following questions:

   i) What do you wonder about after reading the article?
   ii) What did you learn about the process of making a flat map?
   iii) What do you think the author meant by "No flat map of the Earth can be perfect"? Do you agree with this statement? Why or why not?

b) Consider the new two-sided map the author and his colleagues present:

   i) How is it superior to the Mercator map?
   ii) How is it superior to the Winkel tripel map?
   iii) How are the three maps similar or different?

VII. Let's reflect

1) Recently, we have begun to use small screens (such as laptops or cellular phones) to view maps. These maps are usually dynamic. We can zoom in, zoom out, rotate, etc. What are some of the consequences of using these digital maps? What are the advantages and disadvantages of our relatively new system of locating points on Earth and getting information about them?
2) This quest points out the difficulties we face when representing Earth on a piece of paper or other flat surface. We used mathematics as a tool to show how different maps distort reality. In what other contexts can mathematics be used to point out distortions, discrepancies, or inequities?
3) If you recall, Quest 3.1 was about juggling. Compare this quest to Quest 3.1: What is the connection between the mathematics of juggling and the mathematics of cartography? What do they have in common? What elements of creativity and mathematics do they both invoke?
4) How can mapmaking be considered a political activity? Do some research on examples in history when map designs perpetuated inequities.
5) Find a map that you believe intentionally distorts reality. How does it distort reality? What do you think were the biases of the mapmaker?
6) What can be done to address the biases that some maps represent?
7) Recall the features of creativity we listed in Quest 1.1 (see Figure 5.34).

   a) Which of these features of creativity does this quest relate to? Describe how you connected each feature to this quest.
   b) Look over these features. In what ways have you nurtured your creativity throughout this book?
   c) Make your own 'creativi-tree' to represent your perceptions of creativity (see Figure 0.1 in Chapter 0 for reference).

8) This quest also asks you to recall some of the mathematical ways of thinking we employed in the book: Monte Carlo simulations, estimation, and proportional reasoning, for example. How does this quest connect to other mathematics content you previously explored in your journey?

9) For the final task in this book, look back on all of the activities you completed on your journey in the book and reflect on your experiences connecting creativity and mathematics. Answer the following questions and reflect on how your responses compare with those you gave at the end of Chapter 0.

   a) What is my approach to creativity?
   b) What is my approach to mathematics?
   c) What is my approach to creative thinking through mathematics?
   d) How can I continue to connect creativity and mathematics in the future?

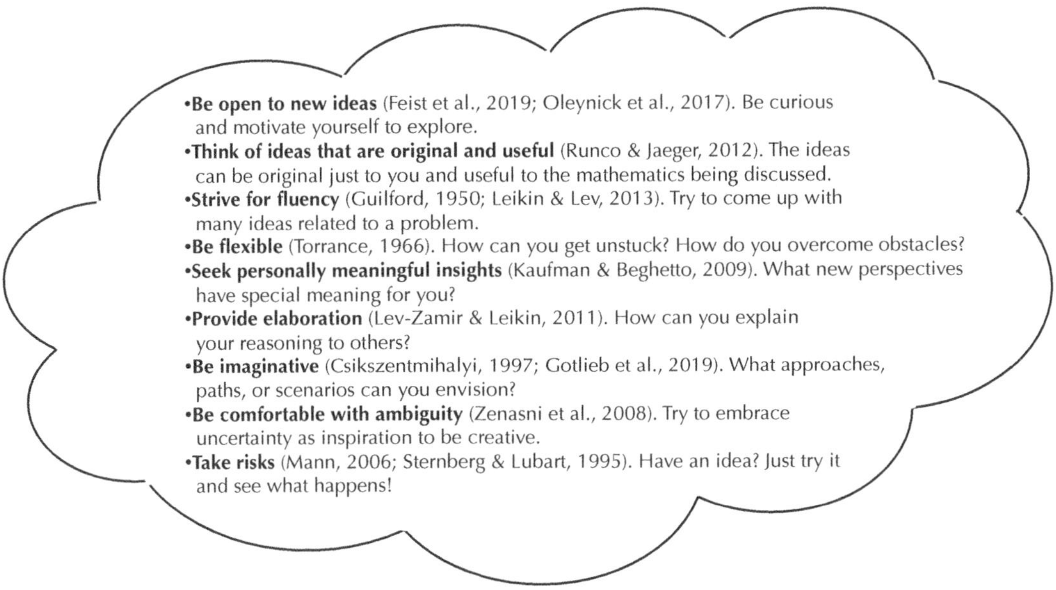

Figure 5.34 Some features of creative thinking, with connections made to mathematics.

Project 5: 'I wonder' log

We frequently find ourselves wondering about the world around us. For example, we see changes in weather patterns and wonder how they will affect the plants growing in our garden. When we see a new spice at a local supermarket, we wonder about ways to use it in our cooking. Often, when we wonder about something, we continue to wonder about it until we find a satisfying solution. Wondering plays an integral part in our lives; why not purposefully use mathematics as we wonder?

In this last project, we encourage you to wonder and to *keep wondering* about things that matter to you using mathematics. Your journey is not over at the end of this book. You, mathematics, and creativity will continue to interact in every facet of your life. What is perhaps most wondrous about mathematics is how mathematical ideas permeate every part of our lives whether we know it or not. Mathematics does not ask anyone for permission to enter their life. It does not wait for an invitation but, rather, just shows up unannounced.

Your present task is to chronicle interesting, un-thought-of-before cases that call for mathematical and creative approaches. You will be noting a variety of experiences and observations that make you wonder. These should be firsthand experiences and observations, as opposed to what you see on TV (or the Internet) or read in a book. Keeping track of these situations will work best if you have a small notebook (or phone) with you to record your musings. Once a suitable problem is identified, you will solve the problem using your creativity and understanding of mathematics.

This project consists of two parts.

Part 1: 'I wonder' log (1 week)

You are going to write in an 'I wonder' log for a week. In this log, you are going to describe situations you encounter that you think will require the use of mathematics. You can jot them down, take pictures, or record audio or video. Note place and time and the problem at hand. These can be framed in the following forms: "Today I noticed that . . . and now I wonder if/how/what . . ." or simply start by writing "I wonder if/how/what . . .". Although there is no limit to the number of situations you note, you should have at least five observations recorded in your log. Read the sample 'I wonder' logs that follow for suggested length, content, and structure of each entry. Try to write each entry as clearly and accurately as possible. Note all relevant information, including details related to the mathematical aspects of each situation.

Here are some examples of entries for Part 1

- *Example 1 – At a restaurant:* When dining at a reservation-only restaurant in town, I noticed how many tables with the 'reserved' placard remained empty. I wonder which model is most cost-effective: a reservation-only restaurant, a no-reservation restaurant, or a restaurant that employs a mix of the two. Certain characteristics of a restaurant might make particular models more beneficial, like popularity, location, or hours open. I wonder what these characteristics are and how one would determine the most cost-effective model for different types of restaurants?
- *Example 2 – Parking meters:* The street meters in our town have been replaced by pay stations. The old meters allowed us to add coins to extend time if our meter was about to expire. If we found a parking spot that still had time on its meter, we got that time for

Figure 5.35 Fish tanks in a college biology laboratory for Example 3. Photograph by Melissa Spigelman.

free. On the other hand, the new pay stations do not allow for this and the new parker cannot use time left over by others. Therefore, the city can easily double-charge if parker #1 pays for more time than necessary and parker #2's time overlaps with the overpaid portion. I wonder how much the city benefits from replacing the old meters with new pay stations.

- *Example 3 – College biology laboratory:* Fish – look at all the fish tanks (see my photograph in Figure 5.35)! The tanks shown hold 300 fish in total. I wonder how you can distribute the fish so that each tank of the same size holds the same number of fish and there are no leftover fish or leftover tanks. Is it possible? Are there possibly multiple solutions?

Part 2: A solution or an attempt at solving one of the questions (1 week)

Choose one of your 'I wonder . . .' ponderings and rephrase it as a mathematical question. Your question might be open-ended or unanswerable. That's okay! Next, either try to find a solution or explain why no solution exists. Your proposed solution may be an actual number or it might take on some other form. If you are developing a model, describe the parameters, what information you would need, and what mathematics would be involved. Your work can be shown on paper, video, PowerPoint, illustration, or another appropriate medium.

Chapter introduction: Cited and related works

Acar, S., Berthiaume, K., & Johnson, R. (2023). What kind of questions do creative people ask? *Journal of Creativity, 33*(3), 100062. https://doi.org/10.1016/j.yjoc.2023.100062

Dweck, C. S. (2006). *Mindset: The new psychology of success*. Random House.

Guilford, J. P. (1950). Creativity. *American Psychologist, 5*(9), 444–454.

Zenasni, F., Besançon, M., & Lubart, T. (2008). Creativity and tolerance of ambiguity: An empirical study. *The Journal of Creative Behavior, 42*(1), 61–73. https://doi.org/10.1002/j.2162-6057.2008.tb01080.x

Quest 5.1: Cited and related works

Gilchrist, A. (1863). *Life of William Blake,"Pictor Ignotus"* (Vol. 2). Macmillan and Company.

Quest 5.2: Cited and related works

Air Miles Calculator. (n.d.). *Air miles calculator*. https://www.airmilescalculator.com/

Euclid. (2006). *The elements* (T. L. Heath, Trans.). Barnes & Noble Publishing. (Original work published ca. 300 BC; Original translation published in 1908).

Quest 5.3: Cited and related works

Gott, J.R. (2021, February 17). The most accurate flat map of Earth yet. *Scientific American*. https://www.scientificamerican.com/article/the-most-accurate-flat-map-of-earth-yet/

List of map projections. (n.d.). *Wikipedia*. https://en.wikipedia.org/wiki/List_of_map_projections

6 Teaching for creativity

This *Teaching for Creativity* chapter is intended for course instructors and other professionals using this book in their courses or programs. For each quest, it includes the objectives and the prerequisite content knowledge required, as well as a list of the relevant creativity traits. The creativity traits are drawn from the list found in Quest 1.1 (see Figure 1.5 and the reference list at the end of Quest 1.1.) The list itself is not meant to be exhaustive, and the traits we present include just some of the traits of creativity we think are the most closely aligned to each individual quest. It is our hope that, as students complete each quest, they come to appreciate and understand the many ways in which they can engage creatively with mathematics, and that, by the end of their journey, they recognize the intersections among these approaches and contribute their own examples that merge creative thinking and mathematics.

This chapter also provides a list of materials and suggested pacing for each quest, based on 100-minute class sessions. A list of recommended readings and electronic resources (such as videos) is included. These resources are suitable for instructors and students alike. For more detailed instructional notes, including printable worksheets and sample student responses, please see our online *Facilitator Guide*, available with the activation code.

Chapter 1 What does it mean to you?

Quest 1.1 Examining our relationship with creativity and mathematics

In this first quest of the journey, students are asked to think about the role of creativity and mathematics in their lives. Students are given several tasks, each of which encourages them to embrace alternative perspectives.

- Prerequisite mathematics understanding
  - Venn diagrams. If any student is unfamiliar with Venn diagrams, it would be worthwhile to review this with the whole class before assigning tasks in the 'Going beyond' section.
- Mathematics explored
  - Finding alternative solutions
  - Spatial reasoning
- Three related characteristics of creativity
  - Be open to new ideas (Feist et al., 2017; Oleynick et al., 2017).
  - Be flexible (Torrance, 1966).
  - Be comfortable with ambiguity (Zenasni et al., 2008).

DOI: 10.4324/9781003541097-7

- Materials
  - For the dots activity:
    - Worksheets with at least four sets of the pairs of dots. These (and all other worksheets mentioned hereafter) are available in the *Facilitator Guide*. Have extra worksheets handy.
  - For the cake problem:
    - Worksheets with circles and tables printed on them
    - Play-doh®
    - Plastic knife or other cutting tool (such as dental floss)
- Pacing
  - One class session
- Additional instructional notes
  - Although this quest is written for students to follow along as they read, it may be best for the quest activities (connecting two points and cutting the cake) to be implemented without the book as an in-class exploration in real time. Otherwise, students will see some 'spoilers' later in the quest.
  - We encourage you to have your students work in groups, especially for the 'What else?' section.
- Recommended reading
  - Boaler, J. (2015). *Mathematical mindsets: Unleashing students' potential through creative math, inspiring messages, and innovative teaching.* Jossey Bass Wiley.
  - Robinson, K. (2011). *Out of our minds: Learning to be creative*. John Wiley & Sons.
  - Su, F. E. (2020). *Mathematics for human flourishing*. Yale University Press.

Quest 1.2 Creative approaches to reasonable estimates

In this quest, after giving an explanation of order of magnitude, we encourage creativity in approaches to estimating answers to various order of magnitude problems. There are two main objectives for students in this quest: the production of reasonable estimates and explanations of the path taken to arrive at those estimates. We have found that, as these types of problems are often open to interpretation and multiple solution paths, they present an opportunity to invite students to engage in dialogue and share their perspectives.

- Prerequisite mathematics understanding
  - An understanding of scaling by 10
  - The questions can be adapted to require more advanced mathematics
- Mathematics explored
  - Number sense
  - Order of magnitude
  - Scientific notation
  - Logarithms

- Three related characteristics of creativity
  - Be comfortable with ambiguity (Zenasni et al., 2008).
  - Provide elaboration (Lev-Zamir & Leikin, 2011).
  - Take risks (Mann, 2006; Sternberg & Lubart, 1995).
- Materials
  - Paper
  - Pen or pencil
  - Calculator (optional)
- Pacing
  - One class session
- Additional instructional notes for the order of magnitude problems in the 'What else?' section
  - We recommend practicing with one of the order of magnitude problems in the 'What else?' section with the entire class before having students work in groups. The Pacific Ocean question elicits many different responses, so it may be a good one to start with.
  - Be sure to tell students that they may not use 'smart' technology to complete the tasks in this quest. The point of the quest will be lost if the students just jump to searching for the solutions online. Allow the use of technology for the purpose of checking the results only after estimates have been made and thoroughly discussed.
  - Give students ample time to discuss the questions posed in the 'What else?' section. We suggest at least 20 minutes.
  - Students should be encouraged to take notes on their process. One option is to assign one note-taker for each question if the students are working in groups. The note-taker would record important information on the different paths (both fruitful and not) that the group took as they attempted each question and keep track of what they did and how they felt when their first idea or path didn't lead to an answer.
  - Students can work individually, in pairs, or in small groups. You could have groups of students working on the same subset of questions and have each report out their estimates and their paths to those estimates. If the actual solution is available, the class can determine which estimates fall within an order of magnitude of the actual value.
  - Adapt the questions to the demographics, location, and interests of the students.
- Recommended reading
  - Arcand, K. K., & Watzke, M. (2017). *Magnitude: The scale of the universe*. Black Dog & Leventhal.
  - Scharf, C. (2017). *The zoomable universe: An epic tour through cosmic scale, from almost everything to nearly nothing*. Scientific American/Farrar, Straus and Giroux.

Quest 1.3 Designing infographics to represent what matters

This quest focuses on the use of infographics to represent data. Students analyze infographics found on the Internet and explore possible pitfalls in their creation. They then create their own infographics to convey a message about an issue they find to be important.

- Prerequisite mathematics understanding
  - Basic knowledge of bar graphs, histograms, pie charts, and scatter plots
  - Descriptive statistics: mean, mode, range, median
- Mathematics explored
  - Similarity
  - Proportions
  - Statistical reasoning
- Three related characteristics of creativity
  - Think of ideas that are original and useful (Runco & Jaeger, 2012).
  - Seek personally meaningful insights (Kaufman & Beghetto, 2009).
  - Provide elaboration (Lev-Zamir & Leikin, 2011).
- Materials
  - Access to the Internet for looking up data and infographics examples
  - Additionally, for creating their infographics
    - Classroom access to a computer with the appropriate software (Adobe, Canva, PowerPoint, or similar) for use in designing an infographic, if they are using technology to create their infographics
    - Poster paper and markers and other craft supplies if creating by hand
- Pacing
  - Two class sessions
- Additional instructional notes
  - The infographics should be not only visually appealing but also clear, accurate, and self-explanatory. Share your expectations for creativity and mathematical rigor with the students and clarify any points of confusion or ambiguity about the purpose of their infographics.
  - Note that Figure 1.11 was designed to be misleading. Students may notice this before they get to the questions asking them about why it is misleading. Encourage this kind of curiosity!
  - Allot time for students to give a short presentation of their infographic and answer questions from peers or have students view each other's infographics during a gallery walk. You can give the students an opportunity to revise their infographic based on the feedback they receive.
  - Encourage students to come up with a visually beautiful graphic – beyond common graphs (for example, bar graphs, scatter plots, histograms, pie charts) – that might illuminate information hidden in the data in an original and attractive way.
- Recommended reading
  - Dick, M. (2020). *The infographic: A history of data graphics in news and communications*. The MIT Press.
  - Tam, K. (2021). *Infographic design: Visual storytelling with information and data*. Gingko Press.
  - Huff, D. (1954). *How to lie with statistics*. W. W. Norton & Company.
  - Schneps, L., & Colmez, C. (2013). *Math on trial: How numbers get used and abused in the courtroom*. Basic Books.

- Tufte, E. R. (1997). *Visual explanations: Images and quantities, evidence and narrative.* Graphics Press.

Chapter 2 What do you notice?

Quest 2.1 Imagining shapes made from a strip of paper

This quest encourages students to imagine the outcome of making loops out of paper strips in different configurations. It enforces the idea that mathematics doesn't necessarily entail only numbers, formulas, or computations.

- Prerequisite mathematics understanding
  - Knowledge of geometric terms
    - Rhombus
    - Parallelogram
    - Diameter
    - Cylinder
  - Basic understanding of the hierarchy of quadrilaterals (for example, a square is a type of rhombus)
- Mathematics explored
  - Geometry
  - Spatial reasoning
  - Measurement
- Three related characteristics of creativity
  - Be open to new ideas (Feist et al., 2017; Oleynick et al., 2017).
  - Take risks (Mann, 2006; Sternberg & Lubart, 1995).
  - Be imaginative (Csikszentmihalyi, 1997; Gotlieb et al., 2019).
- Materials
  - Scissors
  - Multiple sheets of A4 (21 cm by 30 cm) or 8½ inch by 11 inch copier paper: It is best to have three different colors.
  - Tape or glue
  - Ruler
- Pacing
  - One class session (with an additional one for further explorations of the 'Going beyond' section if desired)
  - The 'Going beyond' section includes many variations on the theme. Consider allowing students to select two or three to pursue. Alternatively, assign different problems to groups and have them present their outcomes.
- Additional instructional notes
  - This quest might lead to frustration for some, as things can fall apart (quite literally) if the instructions are not followed carefully or if students are not careful when taping the

  seams. In a day and age when instant gratification is expected, perhaps this is okay. It's a good lesson to slow down, breathe, and start fresh.
- Some configurations require students to cut through the tape connecting two links or two taped short edges of a paper strip. This step might seem odd or wrong. Let the students know that this cut is needed to complete the task.
- Some of the variations in 'What else?' might be difficult for students to imagine. We recommend trying these out first before assigning them to your students.
- For extensions of this quest with a focus on finding areas and perimeters of resulting loops, please see
  Vishnubhotla, M. & Munakata, M. (2018). The mathematics of loopy picture frames. *Teaching Mathematics in the Middle School. 23*(4), 231–234. https://doi.org/10.5951/mathteacmiddscho.23.4.0231

- Recommended readings
  - Heinz, A. (2021). *Folding polyhedra: The art & geometry of paper folding*. Schiffer Craft.
  - Hull, T. C. (2020). *Origametry: Mathematical methods in paper folding*. Cambridge University Press.
  - Lang, R. J. (2011). *Origami design secrets: Mathematical methods for an ancient art* (2nd ed.). A K Peters/CRC Press.

Quest 2.2 Connecting ordinary objects to mathematics

This quest challenges students to consider the role constraints play in promoting creativity. Objects that are seemingly unrelated are 'forced' to be related through connections to mathematics, thus introducing a constraint. The features of the objects and the instructions encourage students to think about how mathematics is related not generally to the world around them but specifically to this limited set of objects, focusing students' attention on particular connections and what the objects represent.

In the next quest, however, students will be given a specific mathematics topic and asked to find representations of that topic in their surroundings. The mathematics will therefore serve as the constraint in Quest 2.3.

- Prerequisite mathematics understanding
  - Any and all previous understandings are useful.
- Mathematics explored
  - Students will identify the mathematics content relevant to the objects.
- Three related characteristics of creativity
  - Strive for fluency (Guilford, 1950; Leikin, 2007).
  - Be flexible (Torrance, 1966).
  - Provide elaboration (Lev-Zamir & Leikin, 2011).
- Materials

  The following are some objects you might provide for the 'What do you notice' task:

  - Empty coffee can
  - Tape
  - Ruler

- Rubber band
- String
- Golf ball
- Toothpick
- Spatula
- Pine cone
- Toothbrush
- Remote control
- Button
- Towel
- Paperclip
- Keys
- Sugar cube
- Tissue paper

- Pacing
  - One session
- Additional instructional notes
  - This quest works best with students working in small groups. We recommend 3–4 students per group.
  - Give students about 45 minutes to collaborate using their tools and to create something mathematical.
  - Consider having students participate in a gallery walk to share and see others' ideas. One student from each group can stay at their station to explain their ideas. After 10 minutes, students should switch roles to allow those that were stationed at their exhibit to view other projects.
- Recommended reading
  - Du Sautoy, M. (2010). *The number mysteries: A mathematical odyssey through everyday life*. St. Martin's Griffin.
  - Ellenberg, J. (2022). *The hidden geometry of information, biology, strategy, democracy, and everything else*. Penguin Books.
  - Fernandez, O. E. (2014). *Everyday calculus: Discovering the hidden math all around us*. Princeton University Press.
  - Paulos, J. A. (2013). *A mathematician reads the newspaper*. Basic Books.

Quest 2.3 Connecting mathematics to ordinary objects

Whereas Quest 2.2 challenges students to draw out the mathematics in a collection of objects, this quest encourages them to view their world with the intention of connecting their surroundings to a few specific mathematics topics. The constraint of having particular mathematics topics in mind focuses the students' attention when taking a tour of their town. The open-ended nature of this quest offers a unique opportunity to get to understand your students' mathematical thinking and prior knowledge. Also, it's a good opportunity for them to go outside and get some fresh air!

- Prerequisite mathematics understanding
  - All previous mathematics understandings are welcome!

- Mathematics explored
  - Students should feel free to select content areas to pursue.
- Three related characteristics of creativity
  - Think of ideas that are original and useful (Runco & Jaeger, 2012).
  - Strive for fluency (Guilford, 1950; Leikin, 2007).
  - Seek personally meaningful insights (Kaufman & Beghetto, 2009).
- Materials for creating their interactive map (in the 'Let's see what happens' section)
  - Measurement tools of the students' choosing.
  - A phone for taking photos or recording audio. A phone can also be used to take notes, but there is something to be said for students using a clipboard, paper, and pen to jot down notes.
  - Access to a laptop or iPad and software such as PowerPoint to make the map. (It'll be even better if students design their maps without electronic devices.)
- Pacing
  - One session
  - Two sessions if following up with a curated math walk
- Additional instructional notes
  - It may not be practical to have students go on the 'math walk' during class. In this case, it can be assigned as an out-of-class activity. (In fact, this may be preferred.)
  - After the maps are constructed, you could dedicate the following session to a curated math walk, where the students view each other's maps and respond to the tasks in them. You could have them discuss their responses with the mapmakers.
- Recommended reading
  - Singh, S. (2013). *The Simpsons and their mathematical secrets*. Bloomsbury Publishing.
  - Strogatz, S. (2012). *The joy of x: A guided tour of math, from one to infinity*. Houghton Mifflin Harcourt.
  - Woo, E. (2019). *It's a numberful world: How math is hiding everywhere . . . from the crown of a tree to the sound of a sine wave*. The Experiment.

Chapter 3 Where is the mathematics in play?

Quest 3.1 Using mathematical notation to make sense of juggling

In this quest, students learn about the mathematics of juggling. Students are encouraged to think creatively as they practice juggling moves and discover mathematical connections through play and patterns.

- Prerequisite mathematics understanding
  - Numerical patterns
  - Multiples and factors

- Mathematics explored
  - Numerical patterns
  - Projectile motion
  - Interpreting diagrams
  - Using mathematics to support one's claims
- Three related characteristics of creativity
  - Be open to new ideas (Feist et al., 2017; Oleynick et al., 2017).
  - Strive for fluency (Guilford, 1950; Leikin, 2007).
  - Be comfortable with ambiguity (Zenasni et al., 2008).
- Materials
  - Four objects for juggling (bean bags, juggling balls, or tennis balls)
  - Colored pencils or pens (optional)
- Pacing
  - Two to three sessions
- Additional instructional notes
  - Juggling should be fun – not stressful. This is a difficult lesson because of the juggling but may be appealing to some because it recognizes and celebrates other talents. The point is to see mathematics in play and support students in discovering some mathematical patterns in juggling.
  - If this activity is done in a larger group setting (for example, a class), we recommend taking some time for students to practice juggling. Make groups of mixed juggling abilities to make sure that students who have mastered juggling are scattered amongst different groups.
  - Let the students develop their own notation first and then come together to analyze one particular notation like the siteswap.
  - This quest can be spread out over several days or weeks to give students time to practice. Consider assigning different juggling patterns, starting with one-ball juggling, throughout the semester.
  - If possible, attend a juggling show or presentation with your students so that they can see professional jugglers and ask them questions about their journey with juggling. Of course, watching videos together is an alternative to this.
- Recommended reading and videos
  - Buhler, J., & Graham, R. (1984). Fountains, showers, and cascades. *The Sciences*, 44–51.
  - Duinker, N. (2013, November 25). *How to juggle 3 balls – tutorial* [Video]. YouTube. https://www.youtube.com/watch?v=whgT2oAIGJc
  - Glenn, T. (2019, September 17). *Learn to juggle 3 balls – beginner tutorial* [Video]. YouTube. https://www.youtube.com/watch?v=dCYDZDIcO6g
  - Naylor, M. (2012). The mathematical art of juggling: using mathematics to predict, describe and create. In *Proceedings of Bridges 2012: Mathematics, Music, Art, Architecture, Culture* (pp. 33–40).

Quest 3.2 Playing with probability using a beachball and finger paint

In this quest, students learn about the Monte Carlo method. They use a simplified version of the Monte Carlo method to calculate relative sizes of areas using finger paint and a beachball. They further explore this method using a map on the ground and using darts and a dartboard.

- Prerequisite mathematics understanding
  - Percentages
  - Ratios
  - Area
- Mathematics explored
  - Monte Carlo method
  - Probability
  - Data analysis
- Three related characteristics of creativity
  - Think of ideas that are original and useful (Runco & Jaeger, 2012).
  - Be comfortable with ambiguity (Zenasni et al., 2008).
  - Take risks (Mann, 2006; Sternberg & Lubart, 1995).
- Materials
  - For tossing the beachball (and globe)
    - An inflatable beachball (a four-color beachball is recommended)
    - Washable finger paint
    - Paper cups (to hold the finger paint)
    - Paper towels (optional for cleaning)
    - Apron (optional for clothes protection)
    - An inflatable globe (found online)
  - A poster-sized map of a geographical area or access to an electronic map and appropriate software. If a large paper map is not available, you can project an image of a map onto poster paper and make a rough outline of the continents. (In fact, that's what we did when we taught this lesson!)
  - A dart board and darts. (If a dartboard is not available, encourage students to make a dartboard out of cardboard or other suitable material.)
- Pacing
  - Two class sessions
- Additional instructional notes
  - In this quest, students use finger paint to track the outcome of tossing a beachball. Before starting the activity, consider giving students some time to practice with the ball to become accustomed to the ball's weight, projectile path, and how it reacts to the physical environment around them. This can improve their ability to collect data later in the quest.
  - This activity can get a bit messy. To minimize excessive splashing or smudging of the ball or another surface (for example, clothing), prepare wet paper towels or wipes for the quest. Also, check the physical space surrounding the students to remove any

possible obstacles in students' way, ensuring that everyone can move around freely. Or better yet, take the students outside to play!
- We recommend that students wear protective aprons over their clothes when using finger paint.
- If students use electronic maps or find an image of a dartboard online, consider discussing how they can randomly generate points on the map or image. For example, they might superimpose a coordinate plane onto the image and use a random number generator to generate an ordered pair to generate a point. In fact, using technology might be preferred or posed as an added challenge.

- Recommended reading
  - Morin, D. J. (2016). *Probability: For the enthusiastic beginner*. CreateSpace Independent Publishing Platform.
  - Patrick, D. (2007). *Introduction to counting and probability* (2nd ed.). AoPS Incorporated.

Quest 3.3 A different take on π with gummy bears and Monte Carlo simulations

In this quest, students apply the Monte Carlo method to derive the value of π using gummy bears distributed randomly inside a circle. They compare estimates of π using a quarter circle and a whole circle.

- Prerequisite mathematics understanding
  - Ratios
  - Fractions
  - Circle, radius, diameter, circumference
  - An understanding of what π is
- Mathematics explored
  - Monte Carlo simulation
  - Probability of an event occurrence
  - Area of a circle
- Three related characteristics of creativity
  - Be open to new ideas (Feist et al., 2017; Oleynick et al., 2017).
  - Provide elaboration (Lev-Zamir & Leikin, 2011).
  - Be comfortable with ambiguity (Zenasni et al., 2008).
- Materials
  - Chalk or masking tape if you will be working outside
  - Large poster paper (about 50 cm by 60 cm) if you will be working inside
  - 5 cups of gummy bears (or, medium size beans or similar)
  - Measuring tape or stick
  - Calculator
  - A marker
- Pacing
  - One to two class sessions

- Additional instructional notes
  - Beans or coins can be used in place of gummy bears. We recommend objects that do not bounce and are easy to count.
  - Collect class data to determine the average estimate for π.
  - For the 'What else' section, we made some decisions about what to alter. Consider other variations and their ramifications on the estimates. The following are some examples of other modifications:
    - Start with a quarter circle inscribed in a 1 meter by 1 meter square, then inscribe a circle in a 2 meter by 2 meter square.
    - Start with an inscribed circle in a square that is 1 meter by 1 meter, then inscribe a circle in a 0.5 meter by 0.5 meter square.
    - Use more or fewer gummy bears.
- Recommended reading
  - Blatner, D. (1999). *The joy of* π. Walker Books.
  - Parker, M. (2021). *Humble Pi: When math goes wrong in the real world.* Riverhead Books.

Chapter 4 Where is the beauty in mathematics?

Quest 4.1 Making art with modular arithmetic

In this quest, students learn about modular arithmetic and where they might encounter it in their everyday lives. In addition, students explore visual representations of modular relationships and design models that reveal the beauty of geometric patterns formed by remainders.

- Prerequisite mathematics understanding
  - Numerical operations
  - Remainders
  - Divisors
  - Evaluating algebraic expressions
- Mathematics explored
  - Numerical patterns
  - Modular arithmetic
  - Number theory
  - Linear relationships
- Three related characteristics of creativity
  - Be flexible (Torrance, 1966).
  - Provide elaboration (Lev-Zamir & Leikin, 2011).
  - Be comfortable with ambiguity (Zenasni et al., 2008).
- Materials
  - An analog 12-hour clock (optional)
  - Calculators (optional)

- Multiple sheets of paper
- Colored pens (recommended)
- Compasses to draw circles

- Pacing
  - Two class sessions
- Additional instructional notes
  - Students may find the notation and the multiple representations of congruent moduli difficult to understand at first. Consider having a whole-class discussion about making sense of the notation and giving students opportunities to practice filling out the tables together. Discuss the meaning of 'mapping' in mathematics.
  - Encourage students to discuss different interpretations of the problems, especially those related to the analog clocks. Some students may want to account for a.m. (morning) or p.m. (afternoon) or stipulate that the 12-hour and 8-hour clocks be synced at a particular time.
  - For questions related to the $kx(mod\,n)$ notation, be sure to have students multiply k by x before determining kx in modulo n. We have intentionally left the kx notation in Table 4.3 so as to not over-scaffold.
  - Transferring the values from the table to the modulo circle may be a challenge. Consider doing part of one together as a class before having groups or individuals map out the rest.
- Recommended reading
  - Maestro, B. (2005). *The story of clocks and calendars.* Harper Trophy.
  - Ornes, S. (2019). *Math art: Truth, beauty, and equations.* Union Square & Co.

Quest 4.2 Exploring recursive and explicit formulas through visual representations

In this quest, students explore patterns of number sequences in many different scenarios. The quest starts with Pascal's triangle, and then goes on to triangular numbers and graphs of the number of selfies between friends, before exploring other visual representations. Students use visual representations of number sequences to develop explicit and recursive formulas and to help them make sense of number patterns in an application problem.

- Prerequisite mathematics understanding
  - Function notation
  - Simplification of algebraic expressions
  - Geometric shapes
- Mathematics explored
  - Number patterns
  - Graph theory
  - Number theory
  - Probability
  - Recursive vs. explicit formulas

- Three related characteristics of creativity
  - Think of ideas that are original and useful (Runco & Jaeger, 2012).
  - Strive for fluency (Guilford, 1950; Leikin, 2007).
  - Seek personally meaningful insights (Kaufman & Beghetto, 2009).
- Materials
  - Multiple copies of printed Pascal's triangle for each student (see *Facilitator Guide* for printable triangles)
  - Multiple copies of diagrams of square and triangular number patterns with open circles
  - A variety of colored pencils, crayons, or markers
- Pacing
  - Three to four class sessions
- Additional instructional notes
  - Encourage students to create their own visual representations of number sequences in Pascal's triangle and to expand upon the visual models presented in the quest.
  - Acting out how the selfies are made with 2, 3, 4, 5, or more people might help students see the patterns in this problem.
  - When developing formulas for triangular numbers, square numbers, or the selfies problem, remind students to look for patterns in the sequence of consecutive values in each type of number.
  - You may need to support students as they consider Questions 27 and 28. Be sure to do these in order. Please refer to the *Facilitator Guide* for sample student responses.
  - There are several extensions of this quest. Some of them (such as probability) are mentioned briefly. You may consider using these extensions as jumping-off points for deeper explorations.
- Recommended reading
  - Maor, E., & Jost, E. (2014). *Beautiful geometry*. Princeton University Press.
  - Suri, M. (2022). *The Big Bang of numbers: How to build the universe using only math*. W. W. Norton & Company.

Quest 4.3 Appreciating self-similarity through infinity

The goal of this quest is to introduce students to the concept of self-similarity and have them apply what they know about self-similarity to create and explore ideas about fractals.

- Prerequisite mathematics understanding
  - Ratios and proportions (scaling)
  - Area and perimeter of a triangle
- Mathematics explored
  - Set notation
  - Geometric sequences

   - Convergence and divergence
   - Self-similarity
   - Classic fractals

- Three related characteristics of creativity

   - Be flexible (Torrance, 1966).
   - Seek personally meaningful insights (Kaufman & Beghetto, 2009).
   - Be imaginative (Csikszentmihalyi, 1997; Gotlieb et al., 2019).

- Materials

   - Printed copies of classic fractals
   - Printed tables for area and perimeter calculations
   - Calculator
   - Broccoli or cauliflower to demonstrate self-similarity
   - Photographs of fractal-like objects in nature

- Pacing

   - Three class sessions

- Additional instructional notes

   - We recommend making sure that students understand the meaning of 'codes' before they proceed on their own. This practice will pay off when students create their own fractals, making and applying their own codes, and moving beyond fractals presented in this quest.
   - As an extension, students can create fractal images that involve rotations (0°, 90°, 180°, and 270°) and reflections (vertical, horizontal, positive diagonal, and negative diagonal). The resulting images will surely 'wow' students while helping them design new types of fractals. For more information about using rotations and reflections in codes for fractals, please see Sobel and Maletsky (1999) and Peitgen et al. (1999).

- Recommended reading

   - Gleick, J. (1988). *Chaos: Making a new science*. Penguin Books.
   - Leszczyński, E., Fernandez, E., & Lahiere, E. (2025). Making fractals: Expanding learning opportunities in the classroom. *LASER Journal*, *3*(1), 1–21.
   - Mandelbrot, B. B. (2012). *The fractalist: Memoir of a scientific maverick*. Pantheon.
   - Mandelbrot, B. B. (1982). *The fractal geometry of nature*. Times Books.
   - Naylor, M. (1999). Exploring fractals in the classroom. *The Mathematics Teacher, 92*(4), 360–366.
   - Peitgen, H.-O., Jürgens, H., Saupe, D., Maletsky, E., & Perciante, T. (1999). *Fractals for the classroom: Strategic activities* (Vol. 3). Springer.
   - Sobel, M. A., & Maletsky, E. M. (1999). *Teaching mathematics: A sourcebook of aids, activities, and strategies* (3rd Edition). Allyn and Bacon.

Chapter 5 What makes you wonder?

Quest 5.1 Surprising encounters with the Möbius strip

In this quest, students learn the mathematics of a Möbius strip by first constructing the original Möbius strip and then introducing variations to it. This quest introduces them to basic ideas from topology and has them predict particular outcomes and provide their reasonings.

- Prerequisite mathematics understanding
  - Unit fractions
  - Length, width, and height
  - Area and perimeter of a rectangle
  - 3D shapes: cylinders and prisms
  - Features of rectangles
- Mathematics explored
  - Geometry
  - Topology
  - Visualization
  - Spatial reasoning
- Three related characteristics of creativity
  - Take risks (Mann, 2006; Sternberg & Lubart, 1995).
  - Be comfortable with ambiguity (Zenasni et al., 2008).
  - Be imaginative (Csikszentmihalyi, 1997; Gotlieb et al., 2019).
- Materials
  - About 15 rectangular paper strips (recommended size of 30 cm by 2.5 cm)
  - Marker or pen
  - Scissors
  - Tape
- Pacing
  - One to two class sessions
- Additional instructional notes
  - The task of creating Möbius strips may take a few attempts, so we recommend having extra paper strips available.
  - Initially, mistakes in cutting along the midline can occur. Remind students to follow the directions in the quest and to use the pictures as guides.
  - When cutting a Möbius strip along its midline, or parallel to it (closer to the edge), students must stay on the appropriate path on the strip without 'crossing over' towards the other edge.
  - When taping together the ends of a paper strip to make a Möbius strip, remind students to tape it so that the seams are completely covered.
- Recommended reading
  - Ernst, B. (1976). *The magic mirror of M.C. Escher.* Random House.
  - Hofstadter, D. R. (1999). *Gödel, Escher, Bach: An eternal golden braid.* Basic Books.

Quest 5.2 Expanding our view of the world through non-Euclidean geometry

This quest encourages students to wonder about the geometry they're familiar with and how it might have to be adapted to account for the curvature of Earth. Having students make connections between previously learned concepts and perspectives encourages them to think creatively as they continue to wonder about mathematics in their surroundings.

- Prerequisite mathematics understanding
  - Measurement of angles with a protractor
  - Basic geometry terminology (for example, point, line, plane, surface, parallel, perpendicular)
  - Pythagorean Theorem
- Mathematics explored
  - Measurement
  - Euclidean geometry
  - Non-Euclidean geometry
  - Geometry of surfaces
  - Ratios and proportions
  - Axiomatic geometry
- Three related characteristics of creativity
  - Be open to new ideas (Feist et al., 2017; Oleynick et al., 2017).
  - Provide elaboration (Lev-Zamir & Leikin, 2011).
  - Take risks (Mann, 2006; Sternberg & Lubart, 1995).
- Materials
  - Print and cut out one paper protractor for each student. Alternatively, print these and bring scissors for students to cut out their own. Printable paper protractors are available online.
  - One beachball for each group of three
  - One paper ruler for each group
  - String, dental floss, or long rubber bands (long enough to wrap around the beachball)
  - Markers to draw on the beachballs. (Erasable markers work best.)
  - Pencil or pen
  - Paper for doodling
  - A globe
- Pacing
  - Two class sessions
- Additional instructional notes
  - We show the flight paths on maps in Figures 5.12–5.14 instead of on globes. This choice leads to questions about distortions inherent in maps, which are related to Quest 5.3. Encourage discussions of this sort.
  - For the beachball activity:
    - This activity works best when students work in groups of three so that there is a third student with clean hands to retrieve the ball if it lands.
    - Ask students to visualize shapes on a sphere before modeling them with physical objects. This is a good opportunity for the students to make and test predictions and visualize objects in their minds.

    - Consider using online interactive tools, like Google Earth, to have students explore other flight paths. Encourage students to first visualize the paths before having them continue with digital tools.
    - To support the learning of the geometry of great circles, consider slicing a soft, round object like an orange or melon to show the circular shape of the cut.
  - For discussions about Euclid's postulates: If possible, give students opportunities to consider each other's interpretations of Euclid's five postulates. There is much to be gained from considering the implications of each postulate and the possibilities for learning from alternative perspectives.
- Recommended reading
  - Mlodinow, L. (2001). *Euclid's window*. Free Press.
  - Skinner, S. (2006). *Sacred geometry: Deciphering the code*. Sterling Publishing.

Quest 5.3 Examining what we value through mapmaking

With this last quest, we end our journey but encourage students to continue wondering about the world around them through the lens of mathematics and creativity. By inviting students to examine various representations of Earth and consider the limitations of each, we strive to nurture their curiosity, critical thinking, and openness to alternative perspectives. Our hope is that they carry this mindset with them into their future endeavors. This final quest references many of the previous quests in the book as a way to have students reflect on their journey.

- Prerequisite mathematics understanding
  - The geometry of shapes on a sphere
  - Ratios and proportions
  - Congruence and similarity
  - Measurement: Area of triangles
- Mathematics explored
  - Geometric projections
  - Scaling
- Three related characteristics of creativity
  - Be open to new ideas (Feist et al., 2017; Oleynick et al., 2017).
  - Be flexible (Torrance, 1966).
  - Be imaginative (Csikszentmihalyi, 1997; Gotlieb et al., 2019).
- Materials
  - Access to sample maps (for example, Mercator, Winkel tripel)
  - Access to online resources (for example, Google Earth)
  - A physical atlas (optional)
  - Tools for mapmaking: paper, markers, rulers, calculators, protractors.
- Pacing
  - One to two class sessions

- Additional instructional notes
  - Creating a new map
    - Designing a new map of the world is not trivial or simple. Support your students as they generate ideas and guide them in their attempts to arrive at their final designs. Let the existing maps guide student work without constraining or limiting their emerging ideas.
    - This quest can be completed individually but we encourage groups of two to three students to allow for debate and discussions, as well as pooling of materials.
  - Tying in with previous quests
    - As a way to tie everything together, students are asked to reflect on their journey. Some of this reflection involves comparing their thinking for this quest to their work on previous quests. We suggest reminding students about these earlier quests and asking them to bring in their earlier work, if applicable.
- Recommended reading

  - Garfield, S. (2013). *On the map: A mind-expanding exploration of the way the world looks*. Avery.
  - Monmonier, M. (2018). *How to lie with maps* (3rd ed.). The University of Chicago Press.

Index

Note: Page numbers in *italics* indicate a figure, and page numbers in bold indicate a table on the corresponding page.